# About this b

This workbook contains practice to support your learning in P5/P6 maths.

Questions split into three levels of increasing difficulty – Challenge 1, Challenge 2 and Challenge 3 – to aid progress.

Symbol to highlight questions that test problem-solving skills.

Total marks boxes for each challenge and topic.

'How am I doing?' checks for self-evaluation.

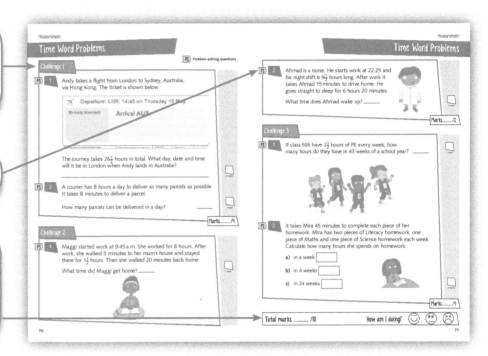

Starter test recaps skills covered in P5/P6.

Four progress tests throughout the book, allowing children to revisit the topics and test how well they have remembered the information.

Progress charts to record results and identify which areas need further revision and practice.

Answers for all the questions are included in a pull-out answer section at the back of the book.

# Contents

# Contents

**ACKNOWLEDGEMENTS**

The author and publisher are grateful to the copyright holders for permission to use quoted materials and images.

All illustrations and images are ©Shutterstock.com and ©HarperCollinsPublishers Ltd

Every effort has been made to trace copyright holders and obtain their permission for the use of copyright material. The author and publisher will gladly receive information enabling them to rectify any error or omission in subsequent editions. All facts are correct at time of going to press.

Published by Leckie
An imprint of HarperCollinsPublishers
Westerhill Road, Glasgow G64 2QT

HarperCollins Publishers
Macken House, 39/40 Mayor Street Upper, Dublin 1
D01 C9W8 Ireland

© 2023 Leckie

ISBN 9780008665906

First published 2017

10 9 8 7 6 5 4 3 2 1

Series Concept and Development: Michelle I'Anson
Commissioning Editor: Richard Toms
Series Editor: Charlotte Christensen
Author: Ali Simpson
Project Manager and Editorial: Tanya Solomons
Cover Design: Sarah Duxbury
Cover Illustration: Louise Forshaw
Inside Concept Design: Ian Wrigley
Text Design and Layout: Contentra Technologies
Artwork: Collins and Contentra Technologies
Production: Natalia Rebow

Printed in United Kingdom

# Starter Test

1. Put these numbers in order from largest to smallest.

| 10.39 | 10.06 | 10.42 | 10.24 | 10.00 |

largest | | | | | | smallest

*1 mark*

2. What is the next term in this sequence?

   4851    4503    4155    3807    _____

   *1 mark*

3. How many triangular faces does a square-based pyramid have? _____

   *1 mark*

4. What is $\frac{1}{2}$ of 8? _____

   *1 mark*

PS 5. Bailey can swim 2 lengths in 3 minutes.
   How long would it take him to swim 6 lengths? _____ minutes

   *1 mark*

PS 6. If Rosalie fills a 2 litre bucket $\frac{1}{5}$ of the way up with water, how much water will she have in the bucket?

   _____ ml

   *1 mark*

**7.** Order these fractions from smallest to largest.

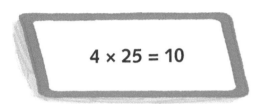

smallest | | | | | largest

1 mark

**8.** Place a decimal point in the correct place to make the calculation correct.

$$4 \times 25 = 10$$

1 mark

**9.** Round Harley's computer score of 8478 to the nearest hundred.

_____

Harley's Score:
8478

1 mark

PS **10.** Lola saves £3.50 from her pocket money every week. How many weeks will it take her to save up to buy her favourite shoes costing £28? _____ weeks

1 mark

**11.** What is $\frac{1}{4}$ of 20? _____

1 mark

# Starter Test

**12.** Look at this timetable.

| Destination | Train 1 | Train 2 | Train 3 |
|---|---|---|---|
| Sopley Bridge | 08:45 | 12:27 | 17:26 |
| Dramford Lane | 09:25 | 13:25 | 18:00 |
| Calshott | 10:02 | 14:06 | 18:24 |
| Inkley Town | 10:56 | 14:42 | 19:37 |
| Travett Centre | 11:29 | 15:16 | 19:58 |
| Somersby | 12:24 | 16:03 | 20:35 |

Immy wants to get from Calshott to Somersby in the fastest time. Which train should she take?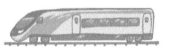

_____

1 mark

**13.** Shade $\frac{3}{4}$ of this shape.

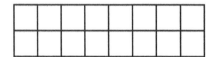

1 mark

**14.** Complete the sequence.

| | |
|---|---|
| 153 278 – 1000 = 152 278 | 152 278 – 1000 = 151 278 |
| 151 278 – 1000 = 150 278 | 150 278 – 1000 = 149 278 |
| _____ – _____ = _____ | _____ – _____ = _____ |

2 marks

**15.** Noah collects football cards. He keeps them in special boxes that each hold 12 cards. How many boxes does he need if he has 283 football cards altogether?

_____ boxes

1 mark

PS **16.** If Connor buys a new water bottle for £6.25 and a lunchbox for £8.99, how much change will he get from £20?    £ _____

**17.** Find a different fraction to put in each box to make the statements correct.

0.75 < ☐

0.75 < ☐

0.75 < ☐

3 marks

**18.** What is the perimeter of this school playground?    _____ m

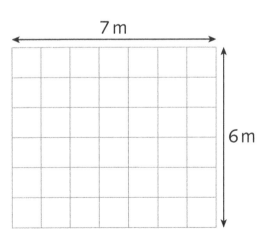

7 m

6 m

1 mark

PS **19.** A bakery is currently running an offer of $\frac{1}{5}$ off all cakes. What will the new price be of a box of cupcakes that were originally priced at £2.50?

£_____

1 mark

**20.** What is $12 \times 7$? _____

1 mark

PS **21.** How many lines of symmetry are there in a pentagon?

_____

1 mark

**22.** Complete these number statements.

a) 84.6 [ ] = 0.846

b) 178.5 [ ] = 17.85

2 marks

**PS** **23.** In the space below, arrange these eight squares to make **three** different shapes, each with a perimeter of 18 cm.

1 cm

1 cm

3 marks

**24.** What is $\frac{2}{5}$ of 210? _____

1 mark

**25.** How many minutes are there in five and a half hours? _____

1 mark

**26.** How many hours are there in $\frac{1}{8}$ of a day? _____

1 mark

**PS** **27.** If Tommy leaves home at 10:10 a.m. and takes a 15 minute walk to the station, then he waits 5 minutes for his train and then takes a 35 minute train ride to work, what time will he arrive at work? _____

1 mark

**PS** **28.** What is the smallest number of equilateral triangles that can make up a regular hexagon? _____

1 mark

**29.** Write a fraction that is equivalent to $\frac{1}{2}$.

1 mark

30. Write these Roman numerals in numbers.

    a) XV _____

    b) XXXIX _____

31. 8593 − 955 = _____

2 marks

1 mark

PS **32.** Mrs Rusling needs 120 chairs set up in the hall in rows, with 8 chairs in each row. How many rows of chairs will there be?

_____

1 mark

PS **33.** Mr Badu is putting up a rectangular fence around the edge of the school playing field. If the perimeter of the field is 76 m, what could the possible dimensions be? _____

1 mark

**34.** Multiply 32 by 30. _____

1 mark

**35.** Write a multiple of 9 between 70 and 80. _____

1 mark

**36.** What fraction of the shape is shaded? _____

1 mark

**37.** Jenni multiplies 86 by 7 and then halves the result. What number is she left with? _____

1 mark

**38.** Calculate $\frac{1}{5}$ of £150.    £_____

1 mark

**39.** Write this number in words.

26 397

_____

_____

1 mark

**PS** **40.** If a ski lesson starts at 10:50 a.m. and lasts for 90 minutes, what time does it finish?

_____

1 mark

Marks........ /47

# Place Value

PS Problem-solving questions

## Challenge 1

**1** Write this number in words.

14 287 _____

_____

1 mark

**2** Write these numbers in order from smallest to largest.

3792    37.92    3710    379.3    3789

_____  _____  _____  _____  _____

1 mark

**3** Complete this sequence.

| 23.9 | ___ | 33.1 | 37.7 | ___ | 46.9 |

2 marks

Marks.......... /4

## Challenge 2

**1** Jayden writes a large number: 39 472.65

What is the value of the following?

**a)** The digit 7 _____

**b)** The digit 5 _____

2 marks

**2** Circle the numbers that have 4 hundreds.

483 296     1 392 482     37 476

284 395     583 406.87

3 marks

# Place Value

**3** If you were to arrange these computer game scores in order, from largest to smallest, whose score would be fifth?

**Petra – 754 109**     **Sammy – 759 184**     **Lilly – 754 119**

**Anisa – 760 367**     **Miguel – 759 185**     **Alexander – 759 439**

_____

1 mark

Marks.......... /6

## Challenge 3

**1** Start with the number **79 496**. Find the number that is:

**a)** 1 more _____     **b)** 10 more _____

**c)** 100 more _____     **d)** 1000 more _____

**e)** 10 000 more _____     **f)** 100 000 more _____

6 marks

**2** If I start with the number 4895, what number do I end with after these operations?

**a)** ÷ 1 _____     **b)** ÷ 10 _____     **c)** ÷ 100 _____

**d)** ÷ 1000 _____

4 marks

**3** Write < or > to make these statements correct.

**a)** 182.29 ☐ 182 290     **b)** –18 ☐ 18

**c)** –37 ☐ –137     **d)** 973 183 ☐ 973 193

4 marks

 **4** Harry starts at 1272. He counts backwards in 163s. What is the fourth number in the sequence?     _____

1 mark

Marks......... /15

Total marks ............. /25     How am I doing?

# Rounding Numbers

## Challenge 1

**1** Which parts of the number change when 683 729 is rounded to the following?

**a)** The nearest 10 _____

**b)** The nearest 1000 _____

*2 marks*

**2** Round these numbers to the nearest thousand.

**a)** 6593 [ ]

**b)** 27 380 [ ]

**c)** 18 730 [ ]

**d)** 329 120 [ ]

**e)** 742 503 [ ]

**f)** 38 294 [ ]

*6 marks*

**3** Round these numbers to the nearest ten thousand.

**a)** 82 390 [ ]

**b)** 29 503 [ ]

**c)** 193 439 [ ]

**d)** 293 663 [ ]

**e)** 275 589 [ ]

**f)** 555 537 [ ]

*6 marks*

**4** Round these numbers to the nearest hundred thousand.

**a)** 283 478 [ ]

**b)** 329 894 [ ]

**c)** 102 038 [ ]

**d)** 482 002 [ ]

**e)** 945 439 [ ]

*5 marks*

Marks......... /19

## Challenge 2

**1** A number rounded to the nearest ten thousand is 30 000.

**a)** What is the largest possible number it could be?

_____

**b)** What is the smallest possible number it could be?

_____

*2 marks*

# Rounding Numbers

**2** Work out the answer to these calculations rounded to 1 decimal place.

a) 22.38 + 92.34 = _____

b) 82.47 + 10.96 = _____

2 marks

Marks.......... /4

## Challenge 3

**1** Round each amount to the nearest pound.

a) £16.38 ☐

b) £20.93 ☐

c) £4.22 ☐

d) £1287.28 ☐

e) £2374.59 ☐

f) £12 629.69 ☐

6 marks

**2** Round these to the nearest whole number.

a) 3.46 ☐

b) 29.85 ☐

c) 36.50 ☐

d) 693.65 ☐

e) 946.74 ☐

f) 47 695.27 ☐

6 marks

**3** Find the answer to each calculation rounded to 1 decimal place.

a) 93.58 + 39.30 = _____

b) 284.23 + 953.58 = _____

2 marks

**4** A newsagent sold 73 672 copies of magazine.

How many copies were sold to the nearest thousand?

_____

1 mark

**5** There are 184 820 ants in a colony.

How many ants are there to the nearest ten thousand? _____

1 mark

Marks.......... /16

Total marks ............. /39

How am I doing?

15

# Roman Numerals

## Challenge 1

**1** Draw a line to match these numbers with their Roman numerals.

| | |
|---|---|
| **1** | XXVIII |
| **5** | XX |
| **10** | C |
| **15** | M |
| **20** | I |
| **28** | LXXXIX |
| **50** | XV |
| **60** | L |
| **64** | X |
| **89** | D |
| **100** | LX |
| **500** | V |
| **1000** | LXIV |

13 marks

Marks......... /13

## Challenge 2

**1** If the Battle of Leicester took place in MCDLV, what year was it? _____

1 mark

**2** If the Queen of Hamilton lost her crown in MDCCCLXXII, what year was it? _____

1 mark

# Roman Numerals

**3** If the pop band The Syndeys sold most albums in MCMXCVI, what year was it? _____  1 mark

**4** In the year MCMLXXXIV Britain got rid of the one pound note. What year was this? _____  1 mark

**5** A new TV will cost the Jones family CMLXXXI pounds. How much money is this? £_____ 1 mark

Marks.......... /5

## Challenge 3

**1** Solve these calculations and write your answers in numbers.

a) IV + XX = _____    b) V × VI = _____

c) XXIII + VIII = _____    d) L + XXV = _____

e) CLIX + XXVI = _____     5 marks

**2** Solve these calculations and write your answers in Roman numerals.

a) LX + L = _____    b) CXCVI ÷ 2 = _____

c) DCVII + CCII = _____    d) M ÷ IV = _____

e) XXX × IX = _____     5 marks

Marks......... /10

Total marks ............ /28    How am I doing?

# Negative Numbers

## Challenge 1

**1** Put these temperatures in order from highest to lowest.

a) 13°C   –4°C   –8°C   8°C   0°C   –10°C

____  ____  ____  ____  ____  ____

b) 0°C   34°C   –19°C   27°C   –2°C   –1°C   28°C   –18°C

____  ____  ____  ____  ____  ____  ____  ____

2 marks

**2** Circle the temperature that is colder.

a) ( 6°C ) or ( –8°C )    b) ( –6°C ) or ( –4°C )

c) ( –28°C ) or ( –29°C )

3 marks

**3** Circle the temperature that is warmer.

a) 12°C or 16°C    b) –1°C or –2°C

c) –8°C or –18°C

3 marks

Marks.......... /8

## Challenge 2

**1** What is the difference between –5°C and 10°C? _____°

1 mark

**2** If the temperature is 3°C and it falls by 14°C,
what is the temperature now?         _____°C

1 mark

**3** What is the difference between 18°C and –9°C? _____°

1 mark

**4** If the temperature is –21°C, how much must
it rise to reach 12°C?               _____°

1 mark

**5** If the temperature rises from 0°C to 8°C
then falls 5°C, what is the new temperature?   _____°C

1 mark

Marks.......... /5

# Negative Numbers

**1** Below is a table showing the average temperatures, measured in degrees, on different planets in different months of the year.

|  | Jan | Feb | Mar | Apr | May | Jun | Jul | Aug | Sep | Oct | Nov | Dec |
|---|---|---|---|---|---|---|---|---|---|---|---|---|
| **Zog** | 9 | 4 | –2 | 5 | 11 | 17 | 1 | 5 | 8 | 10 | 10 | 6 |
| **Silversty** | –20 | –19 | –15 | –11 | –10 | –2 | 0 | –1 | –9 | –13 | –17 | –21 |
| **Kartian** | –5 | 4 | 9 | 13 | 16 | 19 | 20 | 24 | 19 | 11 | 2 | –1 |
| **Sulsbury** | 15 | 16 | 16 | 18 | 21 | 27 | 36 | 39 | 38 | 22 | 14 | 8 |
| **Utophosis** | –18 | –15 | –8 | –2 | 0 | 3 | 5 | 6 | 9 | 1 | –6 | –10 |

**a)** Overall which is the coldest planet? _____

**b)** Overall which is the warmest planet? _____

**c)** What is the difference in temperature between planet Zog and planet Sulsbury in March? _____°

**d)** Which planet had the greatest difference in temperature throughout the year? _____

**e)** Which two months have the greatest difference between the warmest and coldest temperatures on planet Utophosis?

_____

5 marks

Marks.......... /5

Total marks ............. /18    How am I doing?

# Properties of Numbers

PS⟩ **Problem-solving questions**

### Challenge 1

**1**   Write four odd numbers that are multiples of 7.

_____

4 marks

**2**   What are the odd factors of 12? _____

2 marks

PS⟩ **3**   Which two different square numbers under 20
when added together make another square number?

_____

1 mark

Marks............/7

### Challenge 2

PS⟩ **1**   Ally swam a number of lengths of the swimming pool. The
number of lengths is a multiple of 12 and a square number, and
the sum of its digits is 9. Ally swam less than 100 lengths.

How many lengths did Ally swim?        _____

1 mark

**2**   **a)**   What is the smallest 2-digit prime number? _____

**b)**   What is the largest 2-digit prime number? _____

2 marks

# Properties of Numbers

**3** **a)** Complete this factor tree for the number 12.

1 mark

**b)** Draw a factor tree for 48.

1 mark

**c)** Draw two different factor trees for 18.

2 marks

Marks.........../7

## Challenge 3

**1** **a)** $6^2 + 8^2 =$ _____

**b)** $4^2 \times 5^2 =$ _____

**c)** $8^2 \div 2^2 =$ _____

**d)** $12^2 \times 3^2 =$ _____

4 marks

**2** **a)** $3^3 \times 5 =$ _____

**b)** $2^3 + 4^2 =$ _____

**c)** $5^3 - 7^2 =$ _____

**d)** $6^3 \times 2^3 =$ _____

4 marks

Marks.........../8

Total marks ............./22

How am I doing?

# Word Problems

### Challenge 1

**PS** | **1** The winning boat team, *Entrepid*, travelled 310 miles at sea in 4 hours 30 minutes. How long would it take the Entrepid team to travel 3100 miles assuming they were travelling at the same speed? _____ hours

1 mark

**PS** | **2** Emily has 12 princess dolls. For her birthday she is given a large box of new dolls' clothes. In the box are 132 different dolls' dresses. How many different dresses can each doll have if they all have the same amount?

_____

1 mark

**PS** | **3** Marli is making 9 party bags for his friends. He has a total of 56 football cards, 29 erasers and 46 chocolate coins to put in the bags. He wants to share out the items equally. What is the greatest number of each item that Marli can put in the party bags?

_____ football cards    _____ erasers    _____ chocolate coins

3 marks

Marks.......... /5

### Challenge 2

**PS** | **1** If Koah and his brother Zac have 208 marbles altogether, how many do they have each if Zac has a multiple of 12 and Koah has a multiple of 8?

_____

1 mark

**PS** | **2** Miss Staples marks 6 books in half an hour. How many can she mark in 3 hours? _____

1 mark

**PS** | **3** Harry says that there is a number less than 100 that is both a square number and a cube number. Is he right? If so, what is the number? _____

1 mark

Marks.......... /3

# Word Problems

## Challenge 3

PS **1** Christie's age is currently a multiple of 6. Next year her age will be a square number.
How old is Christie? _____ years old

PS **2** If the amount of buttons in Suki's jar is $5^3$ and the amount of buttons in Freddie's jar is $11^2$, who has more buttons in their jar?

_____

PS **3** **a)** Using the information in the table below, work out how many badges each child has collected. Complete the table.

| Name | Description of number | Number of badges collected |
|---|---|---|
| Mae | A square number, multiple of 6, sum of digits is 9, less than 40 | _____ |
| Jake | A prime number, greater than 12, less than 20, sum of digits is 8 | _____ |
| Raheed | A cube number, an odd number, a multiple of 9, less than 30 | _____ |

**b)** Who collected the most badges? _____

PS **4** Complete the table.

| | Answer | Show your working |
|---|---|---|
| **a)** The largest square number under 500 | | |
| **b)** The largest multiple of 4 under 500 | | |
| **c)** The largest cube number under 500 | | |

Marks.......... /9

Total marks ............ /17        How am I doing?

# Addition

**1** Answer these.

  **a)** 2363 + 4820 = _____

  **b)** 6905 + 4795 = _____

  **c)** 4737 + 9952 = _____

  **d)** 9455 + 8764 = _____

4 marks

**2** Answer these.

  **a)** 12 473 + 7239 = _____

  **b)** 48 458 + 4624 = _____

  **c)** 97 421 + 6521 = _____

  **d)** 88 895 + 12 869 = _____

4 marks

**3** Find the missing numbers.

  **a)** 14 457 + ☐ = 19 880

  **b)** 74 404 + ☐ = 83 393

  **c)** 84 330 + ☐ = 101 352

  **d)** 105 576 + ☐ = 129 417

4 marks

Marks.........../12

24

# Addition

## Challenge 2

**1** Use the symbol **<** or **>** to make each statement correct.

a) 48.69 + 39.32 ☐ 54.12 + 32.10

b) 83.15 + 17.23 ☐ 72.46 + 27.99

c) 123.52 + 31.50 ☐ 89.62 + 98.62

d) 321.02 + 129.01 ☐ 294.94 + 89.90

4 marks

Marks.......... /4

## Challenge 3

**1** Answer these.

a) 38.20 + 28.18 + 4.21 = _____

b) 38.98 + 27.40 + 9.27 = _____

c) 337.39 + 128.47 + 3.45 = _____

d) 562.93 + 12.47 + 125.59 = _____

e) 37.42 + 845.35 + 385.93 = _____

5 marks

Marks.......... /5

Total marks ............. /21     How am I doing?

# Subtraction

## Challenge 1

**1** Answer these.

a) 9287 – 2038 = _____

b) 8293 – 2940 = _____

c) 3622 – 2037 = _____

d) 5038 – 4956 = _____

4 marks

**2** Answer these.

a) 29 374 – 8349 = _____

b) 39 023 – 9403 = _____

c) 65 028 – 9482 = _____

d) 94 859 – 28 458 = _____

4 marks

**3** Find the missing numbers.

a) 48 923 – ☐ = 36 138

b) ☐ – 24 812 = 32 168

c) 99 842 – ☐ = 15 627

d) ☐ – 74 206 = 50 481

4 marks

Marks........./12

## Challenge 2

**1** Use the symbol **<** or **>** to make each statement correct.

a) 93.58 – 32.21 ☐ 83.28 – 14.65

1 mark

b) 47.65 – 18.98 ☐ 85.86 – 64.78

1 mark

# Subtraction

c)  234.37 – 90.38 ☐  365.97 – 281.46

d)  832.84 – 233.28 ☐  649.49 – 147.20

Marks.......... /4

## Challenge 3

**1**  Continue the pattern for the next three calculations.

12.03 – 8.98 = 3.05

12.02 – 8.99 = 3.03

3 marks

**2**  Complete the calculation by writing the correct digits in the boxes.

```
    5  4  6  2  ☐
 -  2  ☐  8  3  1
 ─────────────────
    3  0  ☐  9  0
 ─────────────────
```

1 mark

Marks.......... /4

Total marks ............ /20     How am I doing?

# Word Problems

PS Problem-solving questions

## Challenge 1

**PS  1**  In Oak Farm Junior School there is a total of 842 pupils. If there are 233 pupils in Year 3, 182 pupils in Year 4 and 252 pupils in Year 5, how many pupils must be in Year 6?

_____

1 mark

**PS  2**  Jamie is taking part in a team sponsored walk. He has to complete 8 laps of the school field. Then he can tag the next person in his team who does the same number of laps. If there are 114 people in Jamie's team, how many laps in total will they walk?

_____

1 mark

**PS  3**  There are 729 people on a train to Southampton. 68 people get off the train at the first stop and 39 people board the train. How many people are now on the train?

_____

1 mark

Marks.........../3

## Challenge 2

**PS  1**  Four athletes are competing in a charity bike ride. The results table shows the number of metres cycled in the given time each day.

| Athlete | Monday | Tuesday | Wednesday | Thursday | Friday |
|---------|--------|---------|-----------|----------|--------|
| Riya | 23 493 | 10 293 | 22 371 | 9 394 | 42 092 |
| Max | 18 273 | 21 384 | 54 109 | 23 384 | 55 092 |
| Hannah | 10 273 | 53 235 | 33 482 | 18 492 | 56 099 |
| Ashok | 29 302 | 47 038 | 39 272 | 8092 | 9309 |

**a)** Which athlete cycled the greatest distance altogether? _____

1 mark

# Word Problems

**b)** How many more metres did Ashok cycle than Riya?

_____ metres

*1 mark*

**c)** Who cycled the furthest in one day, and on which day was it?

_____

*2 marks*

**d)** On which day was the greatest distance covered by all four cyclists?

_____

*1 mark*

Marks.......... /5

## Challenge 3

**PS** **1** Sanghita's Cafe is open from 9:00 a.m. to 5:00 p.m. every day of the week. Every hour they serve 62 cups of tea and 101 cups of coffee.

**a)** How many teas and coffees are sold altogether in a day?

_____

**b)** How many teas and coffees are sold altogether in a week?

_____

*2 marks*

**PS** **2** The Everton football stadium can hold 40 569 seated spectators. If 3029 of those seats are for away fans, how many seats are for the Everton fans?

_____

*1 mark*

**PS** **3** An ant has 6 legs. Each leg is divided into 3 joints. How many joints are there altogether in 468 ants?

_____

*1 mark*

Marks......... /4

Total marks ............ /12

How am I doing? 😊 😐 😣

# Multiplication

## Challenge 1

**1** Work out the answer to these calculations.

a) 7 × 16 = _____

b) 12 × 19 = _____

c) 32 × 28 = _____

d) 45 × 56 = _____

e) 94 × 82 = _____

5 marks

**2** Work out the answer to these calculations.

a) 119 × 3 = _____

b) 35.2 × 10 = _____

c) 291 × 12 = _____

d) 89.3 × 100 = _____

e) 74.7 × 1000 = _____

5 marks

Marks.........../10

## Challenge 2

**1** Use all the digits **5, 6, 7** and **8** and the × sign once each to make the following:

a) the largest multiplication statement possible

_____

b) the smallest multiplication statement possible

_____

2 marks

**2** Circle the greater amount in each pair.

a) 48 × 24    or    39 × 37

b) 93 × 19    or    84 × 23

c) 183 × 6    or    29 × 38

d) 372 × 18    or    298 × 22

e) 519 × 41    or    846 × 31

5 marks

Marks.........../7

# Multiplication

**Challenge 3**

**1** Write the missing numbers in the boxes to make the calculations correct.

a) $5492 \times 17 =$ [ ]

b) $8324 \times$ [ ] $= 74\,916$

c) [ ] $\times 27 = 28\,053$

d) $3941 \times 45 =$ [ ]

e) $2345 \times$ [ ] $= 21\,105$

5 marks

**2** Make the scales balance by writing the correct value on the other side.

a) $3 \times 12^2$ _____

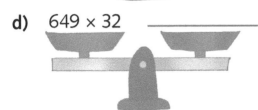

b) $2 \times 16^3$ _____

c) $142 \times 89$ _____

d) $649 \times 32$ _____

4 marks

Marks.......... /9

Total marks ............. /26

How am I doing?

31

# Division

## Challenge 1

**1** Work out the answer to these calculations.

a) 492 ÷ 4 = _____

b) 2688 ÷ 7 = _____

c) 4686 ÷ 6 = _____

d) 12 354 ÷ 3 = _____

e) 1575 ÷ 3 = _____

5 marks

**2** Write a calculation of your own that fits in with this group.

> **620 ÷ 5**
>
> **992 ÷ 8**          **868 ÷ 7**
>
> **372 ÷ 3**
>
> **1116 ÷ 9**

_____

1 mark

Marks.......... /6

## Challenge 2

**1** Assuming you want the most, which of these do you want?

a) 1 cake shared between 5 of you **or** 2 cakes shared between 8 of you

_____

1 mark

b) 14 biscuits shared between 7 of you **or** 30 biscuits shared between 10 of you

_____

1 mark

# Division

c) 24 sweets shared between 3 of you **or** 45 sweets shared between 9 of you

_____

1 mark

d) 42 marshmallows shared between 6 of you **or**
36 marshmallows shared between 9 of you

_____

1 mark

e) 72 jugs of orange juice shared between 8 of you **or** 96 jugs of orange juice shared between 12 of you

_____

1 mark

Marks.......... /5

## Challenge 3

**1** Circle the smaller amount in each pair.

a) $121 \div 11$   or   $72 \div 12$

b) $56 \div 8$   or   $36 \div 6$

c) $81 \div 9$   or   $100 \div 10$

d) $48 \div 6$   or   $56 \div 8$

e) $1230 \div 5$   or   $976 \div 4$

5 marks

**2** Write three different division calculations that each equal 28.

_____   _____   _____

3 marks

**3** Write three different division calculations that each have a remainder of 1.

_____   _____   _____

3 marks

Marks.......... /11

Total marks ............ /22        How am I doing?

# Word Problems

## Challenge 1

**PS** **1** Lissa and Niall have been collecting shells. Altogether they have 187 shells. Lissa has 23 more shells than Niall. How many shells do they each have?

Lissa = _____          Niall = _____

2 marks

**PS** **2** A builder is laying out his bricks. He needs enough bricks to build a wall that is 162 cm long. Each brick is 9 cm long. How many bricks does the builder need to build the first row of the wall? _____

1 mark

**PS** **3**  Mr Campbell wants to plant rose bushes around the outside of his garden. Each rose bush is 1 m wide and he wants to plant the rose bushes 1.5 m apart. If the perimeter of his garden is 20 m, how many rose bushes will Mr Campbell need? _____

1 mark

Marks.......... /4

## Challenge 2

**PS** **1** Two types of alien live on the planet Starlight: Zoomas have 3 eyes and 2 legs and Popsies have 4 eyes and 5 legs. How many of each type of alien might there be if there are 38 eyes and 37 legs?

_____

1 mark

**PS** **2** Michael works for the road safety team on the motorway. He has to lay out a cone every 90 cm for 2520 cm to close a lane.

How many cones will Michael need altogether? _____

1 mark

# Word Problems

PS 3 Miss Cavan is setting up a cricket match for her year group. She wants to make sure that 6 people in each class have a cricket bat at all times. If there is a total of 4 classes, each with 33 pupils in them, how many cricket bats will Miss Cavan need altogether?

_____ 1 mark

Marks.........../3

## Challenge 3

PS 1 Karen runs a jewellers and is making some silver pendants. She has a piece of silver that is 420 cm long and each pendant is made out of 8 cm of silver. How many pendants can Karen make out of one piece of silver?

_____ 1 mark

PS 2 To make a banana and strawberry smoothie, Josie needs one and a half bananas and 8 strawberries, 2 scoops of ice cream and 250 ml of milk. How much of each ingredient will Josie need if she is making a smoothie for each of her 9 friends?

_____ bananas, _____ strawberries,

_____ scoops of ice cream and _____ ml of milk

4 marks

PS 3 David goes on a 90 minute bike ride three times a week. How long will he cycle for in minutes in these times?

a) 2 weeks [ ] minutes

b) 6 weeks [ ] minutes

c) 9 weeks [ ] minutes

3 marks

Marks.........../8

Total marks ............ /15        How am I doing?   ☺  😐  😣

35

1. What is the value of the digit 4 in 845 923?

   _____

   1 mark

2. What is 35 249 rounded to the nearest 100?

   _____

   1 mark

PS 3. Write the next two numbers in this sequence.

| 15.12 | 11.64 | 8.16 | ___ | ___ |

2 marks

PS 4. If Joey receives £4.50 pocket money a week and he always saves half of it, how much money will Joey have saved after 12 weeks?

   £_____

   1 mark

5. Add together the factors of 12.

   _____

   1 mark

6. Round each amount to the nearest pound.

   a) 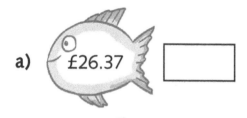 £26.37

   b)  £8.94

   c)  £103.58

   3 marks

PS 7. Complete the sequence by finding the missing number.

   1    4    9        25    36

   1 mark

36

**8.** Assuming you want the largest amount, would you rather have 36 cupcakes shared between 6 of you or 105 cupcakes shared between 15 of you?

_____

**9.** Write < or > to make this statement correct.

483 940 ☐ 48 394.02

**10.** 4140 ÷ 3 = _____

**11.** Write the roman numeral MCMXXXVI in numbers.

_____

**12.** 84 402 – 49 623 = _____

**13.** Complete this sequence by filling in the missing numbers:

121 | ____ | ____ | 25 | 9 | 1

**14.** What is the difference between 24°C and –5°C?  _____

**15.** Calculate 3.7 × 8.

_____

**16.** What is $9^3$?  _____

**17.** Calculate 495 539 – 382 758.

_____

1 mark (×9)

 **PS** Problem-solving questions

**18.** Write < or > to make this statement correct.

10 045.68 ☐ 10 485.68

*1 mark*

**19.** What is the year MDCCXCIX?

_____

*1 mark*

**20.** If the temperature is −17°C, by how much must it rise to reach 31°C?

_____°C

*1 mark*

**21.** Draw a factor tree for the number 40.

*1 mark*

**22.** $10^3 + 8^3 =$ _____

*1 mark*

**23.** Circle the greater amount.

$84 \times 272$   **or**   $92 \times 219$

**PS** **24.** If there are 72 red aliens on Planet Zupta each with 4 legs and 6 eyes, how many legs and eyes are there altogether?

_____

**25.** Fill in the missing number.

$82\,738 + \boxed{\phantom{00000}} = 128\,710$

Marks........./28

# Equivalent Fractions

## Challenge 1

**1** What fraction of the shape is shaded?

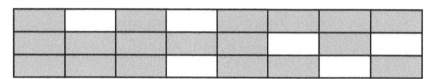

**2** What fraction of the shape is shaded?

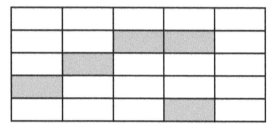

**3** What fraction of the shape is shaded?

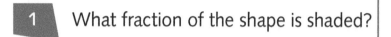

**4** Shade $\frac{1}{3}$ of the shape.

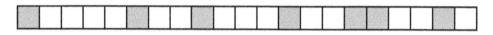

**5** Shade $\frac{2}{10}$ of the shape.

Marks.......... /5

## Challenge 2

**1** Use the denominators given to write equivalent fractions for $\frac{1}{3}$.

 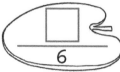 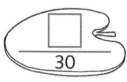 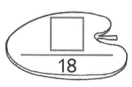

# Equivalent Fractions

**2** Write three equivalent fractions for $\frac{2}{3}$.

□ □ □

3 marks

**3** Write three equivalent fractions for $\frac{1}{5}$.

□ □ □

3 marks

Marks......... /10

## Challenge 3

**1** The fractions in each set are all equivalent fractions, except for one. Circle the odd one out in each set.

a)  $\frac{8}{32}$   $\frac{7}{28}$   $\frac{1}{4}$   $\frac{3}{8}$   $\frac{6}{24}$

b)  $\frac{2}{10}$   $\frac{3}{15}$   $\frac{4}{20}$   $\frac{6}{30}$   $\frac{8}{50}$

c)  $\frac{10}{45}$   $\frac{2}{9}$   $\frac{20}{90}$   $\frac{6}{26}$   $\frac{40}{180}$

d)  $\frac{6}{14}$   $\frac{30}{70}$   $\frac{18}{42}$   $\frac{9}{27}$   $\frac{15}{35}$

4 marks

Marks......... /4

Total marks ............. /19     How am I doing?

41

# Improper Fractions

**1** Draw a line to match each improper fraction to its mixed number equivalent.

 $4\frac{3}{5}$

 $8\frac{2}{10}$

 $\frac{7}{3}$

 $\frac{8}{4}$

 $6\frac{4}{5}$

 $\frac{10}{4}$

 $\frac{74}{9}$

$\frac{34}{5}$

$8\frac{2}{9}$

$2\frac{1}{3}$

$\frac{82}{10}$

$\frac{23}{5}$

2

$2\frac{2}{4}$

7 marks

Marks.........../7

42

# Improper Fractions

## Challenge 2

**1** Convert these mixed numbers into improper fractions.

a) $1\frac{4}{5}$ ☐

b) $2\frac{8}{10}$ ☐

c) $4\frac{5}{9}$ ☐

d) $8\frac{1}{2}$ ☐

e) $6\frac{4}{6}$ ☐

f) $12\frac{11}{12}$ ☐

6 marks

**2** Convert these improper fractions into mixed numbers.

a) $\frac{4}{3}$ ☐

b) $\frac{15}{10}$ ☐

c) $\frac{24}{10}$ ☐

d) $\frac{49}{8}$ ☐

e) $\frac{35}{6}$ ☐

f) $\frac{89}{9}$ ☐

6 marks

Marks.........../12

## Challenge 3

**1** Write $\frac{290}{9}$ as a mixed number. ☐

1 mark

**2** Write $67\frac{1}{6}$ as an improper fraction. ☐

1 mark

Marks.........../2

Total marks ............. /21     How am I doing? 🙂 😐 😣

# Comparing and Ordering Fractions

## Challenge 1

**1** Write < or > to make each statement correct.

a) $\frac{3}{6}$ ☐ $\frac{5}{6}$    b) $\frac{8}{10}$ ☐ $\frac{9}{10}$

c) $\frac{1}{5}$ ☐ $\frac{2}{5}$    d) $\frac{7}{8}$ ☐ $\frac{5}{8}$

4 marks

Marks.......... /4

## Challenge 2

**1** Put these fractions in order, from smallest to largest.

$\frac{1}{3}$    $\frac{1}{2}$    $\frac{1}{5}$    $\frac{1}{4}$    $\frac{1}{6}$    $\frac{1}{10}$

smallest ☐ ☐ ☐ ☐ ☐ ☐ largest

1 mark

**2** Put these fractions in order, from smallest to largest.

$5\frac{2}{6}$    $5\frac{2}{26}$    $5\frac{2}{10}$    $5\frac{1}{2}$    $5\frac{1}{4}$

smallest ☐ ☐ ☐ ☐ ☐ largest

1 mark

**3** Put these fractions in order, from largest to smallest.

$10\frac{1}{3}$    $10\frac{3}{6}$    $10\frac{4}{9}$    $10\frac{12}{12}$    $10\frac{5}{6}$

largest ☐ ☐ ☐ ☐ ☐ smallest

1 mark

Marks.......... /3

# Comparing and Ordering Fractions

**1** Assuming you want the **largest** portion of chocolate cake, which would you rather have?

a)  or

b) $\frac{5}{6}$ or $\frac{2}{3}$

c) $\frac{4}{15}$ or $\frac{3}{10}$

d) $\frac{3}{4}$ or $\frac{16}{20}$

e) $\frac{1}{3}$ or $\frac{2}{9}$

5 marks

**2** Assuming you want the **smallest** portion of pizza, which would you rather have?

a) $\frac{3}{5}$ or $\frac{4}{10}$

b) $1\frac{2}{6}$ or $\frac{9}{6}$

c) $\frac{10}{5}$ or $\frac{7}{3}$

d) $2\frac{3}{4}$ or $\frac{9}{4}$

e) $4\frac{1}{3}$ or $\frac{12}{3}$

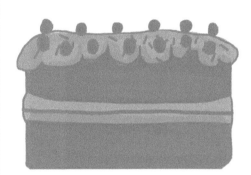

5 marks

Marks.........../10

Total marks ............/17      How am I doing?

45

# Adding and Subtracting Fractions

## Challenge 1

**1** Complete these fraction calculations.

a) $\frac{1}{4} + \frac{1}{4} =$ 

b) $\frac{3}{5} + \frac{2}{5} =$ 

c) $\frac{8}{10} + \frac{1}{10} =$ 

d) $\frac{9}{12} - \frac{2}{12} =$ 

e) $2 - \frac{1}{4} =$ 

f) $\frac{2}{3} - \frac{1}{2} =$ 

g) $\frac{10}{12} - \frac{3}{12} =$ 

h) $\frac{3}{8} - \frac{1}{4} =$ 

8 marks

Marks.......... /8

## Challenge 2

**1** Write your answer to each calculation as an improper fraction.

a) $\frac{1}{3} + \frac{2}{3} =$ 

b) $\frac{6}{8} + 1\frac{2}{8} =$ 

c) $1\frac{3}{4} + \frac{2}{4} =$ 

d) $2\frac{3}{5} - 1\frac{1}{5} =$ 

e) $4\frac{2}{8} - 2\frac{3}{8} =$ 

f) $2\frac{6}{10} + \frac{4}{10} =$ 

g) $4\frac{3}{7} - 1\frac{2}{7} =$ 

h) $8\frac{2}{6} + 9\frac{1}{6} =$ 

8 marks

Marks.......... /8

46

# Adding and Subtracting Fractions

**Challenge 3**

**1** Write your answer to each calculation as a mixed number.

a) $\frac{3}{5} + \frac{4}{5}$ = ☐

b) $\frac{8}{10} + \frac{9}{10}$ = ☐

c) $\frac{10}{20} - \frac{8}{20}$ = ☐

d) $\frac{3}{7} + 2\frac{5}{7}$ = ☐

e) $1\frac{5}{6} - \frac{4}{6}$ = ☐

f) $\frac{5}{8} + 2\frac{4}{8}$ = ☐

g) $4\frac{1}{7} + 1\frac{2}{7}$ = ☐

h) $8\frac{1}{2} - \frac{3}{4}$ = ☐

8 marks

Marks.......... /8

Total marks ............. /24

How am I doing?

# Rounding Decimals

## Challenge 1

**1** Round these numbers to the nearest whole number.

a) 7.18

b) 26.85

c) 33.54

d) 102.93

e) 5220.40

f) 39 490.64

g) 54 947.36

h) 739 943.98

i) 932 555.55

9 marks

Marks.......... /9

## Challenge 2

**1** Round these numbers to the nearest tenth.

a) 16.83

b) 29.48

c) 73.56

d) 283.54

e) 1304.62

f) 49 024.88

6 marks

Marks.......... /6

48

# Rounding Decimals

**Challenge 3**

**1** Which dog is the odd one out? Using rounding, explain your answer.

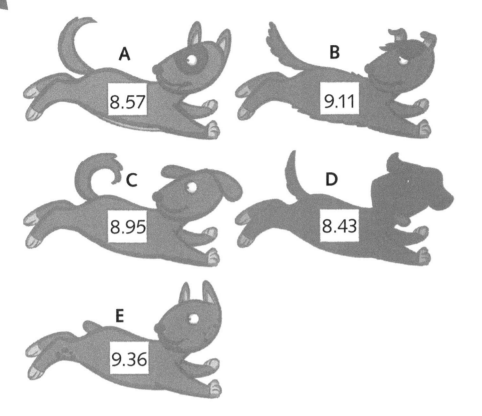

**A** 8.57

**B** 9.11

**C** 8.95

**D** 8.43

**E** 9.36

2 marks

**2** Round these numbers to the nearest tenth.

a) 7.33

b) 9.15

c) 83.29

d) 273.48

e) 827.32

f) 2358.63

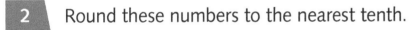

6 marks

Marks.......... /8

Total marks ............. /23

How am I doing?

# Fraction and Decimal Equivalents

## Challenge 1

**1** Draw a line to match each decimal to its fraction equivalent.

| Decimals | Fractions |
|----------|-----------|
| 0.5 | $\frac{7}{20}$ |
| 0.75 | $\frac{9}{10}$ |
| 0.8 | $\frac{63}{100}$ |
| 0.12 | $\frac{1}{2}$ |
| 0.35 | $\frac{3}{4}$ |
| 0.63 | $\frac{3}{25}$ |
| 0.9 | $\frac{4}{5}$ |

7 marks

Marks.........../7

50

# Fraction and Decimal Equivalents

## Challenge 2

**1** Write the decimal fraction for each fraction.

**a)** 1 quarter [ ]   **b)** 3 quarters [ ]

**c)** 2 halves [ ]   **d)** 2 thirds [ ]

**e)** 4 tenths [ ]   **f)** 5 tenths [ ]

6 marks

**2** Write the decimal fraction for each fraction.

**a)** $\frac{2}{8}$ _____   **b)** $\frac{1}{5}$ _____

**c)** $\frac{3}{3}$ _____

3 marks

Marks.......... /9

## Challenge 3

**1** Write < or > to make each statement correct.

**a)** 0.51 [ ] $\frac{1}{2}$   **b)** 0.7 [ ] $\frac{8}{10}$

**c)** 3.75 [ ] $3\frac{2}{4}$   **d)** 0.5 [ ] $\frac{6}{10}$

**e)** 0.84 [ ] $\frac{8}{10}$   **f)** 2.39 [ ] $2\frac{3}{10}$

6 marks

**2** Fill in a missing decimal number to make each statement correct.

**a)** _____ < 2 tenths   **b)** $\frac{5}{15}$ > _____

**c)** _____ < $\frac{3}{4}$

3 marks

Marks.......... /9

Total marks ............ /25   How am I doing?

51

# Fraction, Decimal and Percentage Equivalents

## Challenge 1

**1** The stars below contain **eight** equivalent fraction, decimal and percentage pairs. Use different coloured pencils to match the pairs.

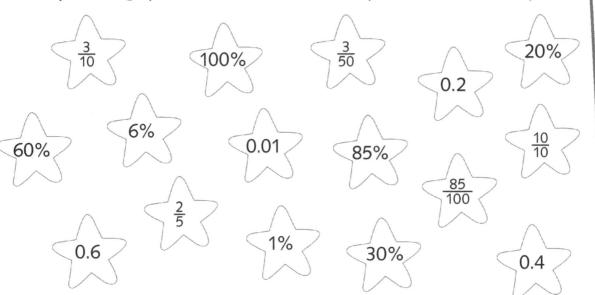

$\frac{3}{10}$  100%  $\frac{3}{50}$  20%  0.2

60%  6%  0.01  85%  $\frac{10}{10}$

$\frac{2}{5}$  $\frac{85}{100}$

0.6  1%  30%  0.4

8 marks

Marks.......... /8

## Challenge 2

**1** Circle the greater amount in each pair.

a) $\frac{1}{2}$    or    60%

b) $\frac{1}{4}$    or    20%

c) $\frac{9}{10}$    or    99%

d) $\frac{1}{100}$    or    10%

e) $\frac{6}{8}$    or    68%

f) $\frac{2}{5}$    or    35%

6 marks

Marks.......... /6

# Fraction, Decimal and Percentage Equivalents

## Challenge 3

**1** Which would you rather have?

a) | 50% of £250 | or | $\frac{1}{4}$ of £460 | _____

b) | 0.1 of £80 | or | $\frac{3}{10}$ of £20 | _____

c) | 25% of £1000 | or | $\frac{4}{5}$ of £700 | _____

d) | 0.4 of £860 | or | 60% of £400 | _____

e) | 20% of £2700 | or | $\frac{2}{5}$ of £1450 | _____

f) | $\frac{3}{4}$ of £100 500 | or | 0.5 of £150 800 | _____

6 marks

**2** A showroom is selling campervans in a sale. Tick the best offer.

75% off £9834          0.8 off £8961          $\frac{9}{10}$ off £9156

1 mark

Marks.......... /7

Total marks ............. /21          How am I doing?   😊   😐   😣

# Multiplying Fractions

## Challenge 1

**1** Answer these. **Example:** $\frac{1}{2} \times 5 = \frac{1 \times 5}{2} = \boxed{\frac{5}{2}}$ **or** simplified to $2\frac{1}{2}$

a) $\frac{1}{3} \times 2 =$ 

b) $\frac{1}{4} \times 8 =$ 

c) $\frac{2}{8} \times 4 =$ 

d) $\frac{3}{5} \times 5 =$ 

e) $\frac{6}{10} \times 10 =$ 

f) $\frac{4}{5} \times 10 =$ 

g) $\frac{8}{9} \times 9 =$ 

7 marks

Marks.........../7

## Challenge 2

**1** Answer these.

**Example:** $2\frac{4}{8} \times 3$

$$2\frac{4}{8} \times 3 = \frac{20}{8} \times 3 = \frac{20 \times 3}{8} = \frac{60}{8} = \frac{15}{2}$$

a) $2\frac{3}{4} \times 5$

1 mark

b) $3\frac{1}{6} \times 3$

1 mark

c) $9\frac{4}{5} \times 4$

1 mark

# Multiplying Fractions

**d)** $8\frac{1}{2} \times 6$

1 mark

**e)** $6\frac{2}{3} \times 2$

1 mark

Marks.......... /5

## Challenge 3

**1** Write **<** or **>** to make each statement correct.

**a)** $\frac{1}{3} \times 4$ ☐ $\frac{5}{3} \times 2$

**b)** $\frac{5}{12} \times 6$ ☐ $\frac{7}{12} \times 5$

**c)** $\frac{7}{7} \times 7$ ☐ $\frac{9}{7} \times 5$

**d)** $\frac{8}{10} \times 6$ ☐ $\frac{5}{10} \times 11$

**e)** $\frac{2}{16} \times 9$ ☐ $\frac{4}{16} \times 4$

**f)** $\frac{9}{22} \times 11$ ☐ $\frac{10}{22} \times 10$

6 marks

Marks......... /6

Total marks ............. /18          How am I doing? 😊 😐 😣

# Word Problems

## Challenge 1

PS **1** Iago's sticker album needs 832 stickers to fill it. So far, Iago has collected 75% of the stickers. How many more stickers does he need to fill his album?

_____

*1 mark*

PS **2** Ethan has raised 60% of the money he needs to pay for his school trip. What percentage of the money does he still need to raise?

_____ %

*1 mark*

Marks........../2

## Challenge 2

PS **1** The Water's Edge restaurant is currently running an offer of 15% off all food.

Chicken Curry ................ £6.20

Onion Soup ................... £4.00

Naan Bread ..................... £2.80

Mint Parfait .................. £3.80

Ice Cream Delight .......... £2.60

**Now with 15% off every item!**

How much do Dawn and Martin spend if they buy 2 chicken curries, 2 onion soups, 1 naan bread, 1 mint parfait and 1 ice cream delight?

£_____

*1 mark*

# Word Problems

PS 2 Jethro is looking at a tablet priced £282. How much would Jethro save if he bought 3 tablets for work with a discount of $\frac{1}{4}$ off the original price?

£_____

£282

1 mark

Marks............/2

## Challenge 3

PS 1 Luke wants to buy a TV. He has seen the TV that he wants in two different shops. TVs R Us are selling the TV for £1984 but with a 30% discount. Electronic Deals are selling the TV for the same price but with a 25% discount. What fraction discount is each shop offering?

TVs R Us = _____        Electronic Deals = _____

2 marks

PS 2 a) Stefan is 180 cm tall. His best friend is $\frac{6}{8}$ of his height.

How tall is his best friend? _____ cm

b) There are 30 pupils in Foxes Class. $\frac{2}{3}$ of pupils walk to school.

How many pupils don't walk to school? _____

2 marks

Marks............/4

Total marks ............/8        How am I doing?

1. Calculate 9347 − 468. _____

1 mark

2. What fraction of the shape is **unshaded**? ☐

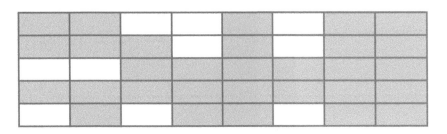

1 mark

3. Write $2\frac{6}{8}$ as an improper fraction. ☐

1 mark

4. Use the denominators given to write equivalent fractions for $\frac{3}{4}$.

$\dfrac{\phantom{0}}{12}$    $\dfrac{\phantom{0}}{8}$    $\dfrac{\phantom{0}}{40}$    $\dfrac{\phantom{0}}{36}$

4 marks

5. Write three equivalent fractions for $\frac{2}{5}$.

3 marks

6. Put these fractions in order from smallest to largest.

$\dfrac{4}{8}$    $\dfrac{3}{4}$    $\dfrac{6}{6}$    $\dfrac{4}{16}$    $\dfrac{10}{12}$

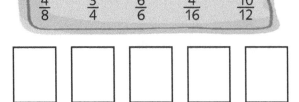

1 mark

7. Round 74 857:

   a) to the nearest ten _____

   b) to the nearest hundred _____

2 marks

8. $\dfrac{5}{12} + \dfrac{4}{12} =$ ☐

1 mark

**9.** $4\frac{3}{9} + 3\frac{8}{9} =$ ☐

**10.** Write $\frac{37}{12}$ as a mixed number. ☐

**11.** $1\frac{3}{5} - \frac{4}{5} =$ ☐

**12.** $2\frac{2}{4} \times 8 =$ ☐

**13.** Complete the sequence by finding the missing numbers. Explain the rule.

10 824    8349    _____    3399    _____

_____

**14.** Look at the adverts for two shops below. Which shop is offering the best saving?

Shelby's – Was: £394
Now: £280

Duson's – Was £176
Now: $\frac{1}{5}$ off

_____

**15.** Write $\frac{2}{5}$ as a decimal. _____

**16.** Write three equivalent fractions for $\frac{1}{4}$.

☐  ☐  ☐

**17.** Round 136.39 to the nearest whole number. _____

**Problem-solving questions**

**18.** Round 632.87 to the nearest tenth. _____

1 mark

**19.** Write 1.5 as a fraction. ☐

1 mark

**20.** Circle the smaller amount.

0.8 of 70   **or**   60% of 90

1 mark

**21.** Put these amounts in order from largest to smallest.

$\frac{1}{5}$ of 886          0.5 of 364

25% of 660          60% of 310

_____

_____

1 mark

**22.** Shade $\frac{2}{5}$ of this shape.

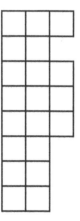

1 mark

**23.** What is the value of 3 in the number 534 879? _____

1 mark

**24.** Write 20% as a fraction. ☐

1 mark

**25.** $5\frac{1}{3} \times 3 =$ ☐

1 mark

**26.** What is the year MCMXCVII in numerals? _____

1 mark

**27.** What is the value of the 5 in the number 17 534 186?

_____

1 mark

**28.** Calculate 1428 × 24.

_____

1 mark

**29.** Subtract 74 035 from 190 954.

_____

1 mark

**30.** Calcalate 2465 shared between 12. Round your answer to the nearest 10.

_____

1 mark

**31.** Draw a factor tree for 525.

1 mark

**32.** On a hot day the temperature in the classroom is 26°C. Mr Freeman turns on the air con and after a short time the temperature has dropped by 11°C. What is the temperature in the classroom now?

_____°C

1 mark

Marks........ /42

## Comparing Measures

**1**  What are these measurements in centimetres?

a)  4 m _____

b)  6.5 m _____

c)  10 m _____

d)  $20\frac{1}{4}$ m _____

e)  5000 m _____

f)  10 750 m _____

6 marks

**2** Convert these measurements into the required units.

a)  2 litres = _____ ml

b)  0.6 kg = _____ g

c)  3100 m = _____ km

d)  74 cm = _____ m

e)  5250 ml = _____ litres

f)  6110 g = _____ kg

6 marks

Marks.........../12

**1**  Write < or > to make these statements correct.

a)  394 ml ☐ 0.4 litres

b)  6.3 km ☐ 630 m

c)  192 cm ☐ 1090 mm

d)  808 g ☐ 0.8 kg

e)  20 956 ml ☐ 209 litres

5 marks

Marks.........../5

# Comparing Measures

## Challenge 3

**1** Circle the heaviest sack of potatoes.

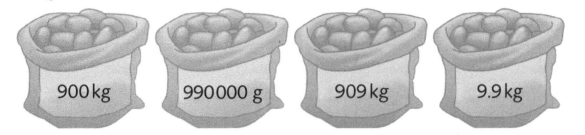

900 kg    990000 g    909 kg    9.9 kg

1 mark

**2** Write these capacities in descending order.

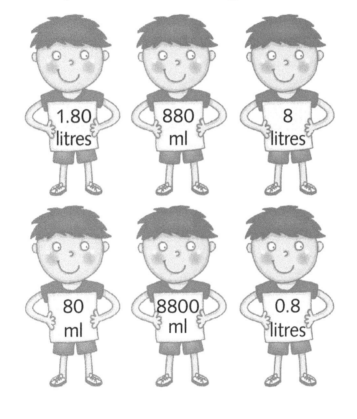

1.80 litres    880 ml    8 litres

80 ml    8800 ml    0.8 litres

_____

_____

1 mark

Marks.........../2

Total marks ............ /19          How am I doing?

# Perimeter

## Challenge 1

**1** Use all 10 of these squares to draw one shape with a perimeter of 22 cm.

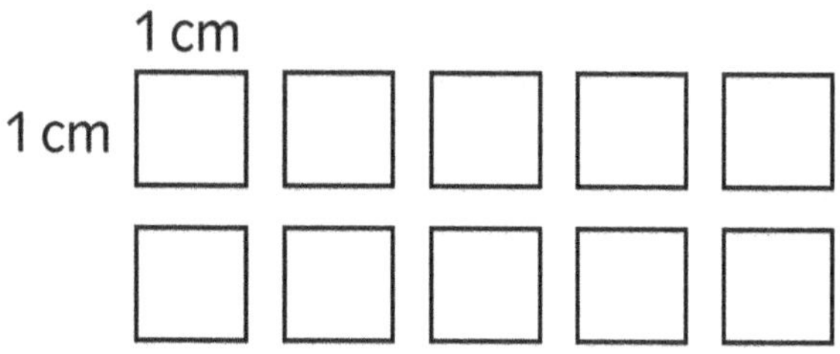

1 cm

1 cm

1 mark

**2** Draw four different shapes, each with a perimeter of 14 cm, using all 6 squares each time.

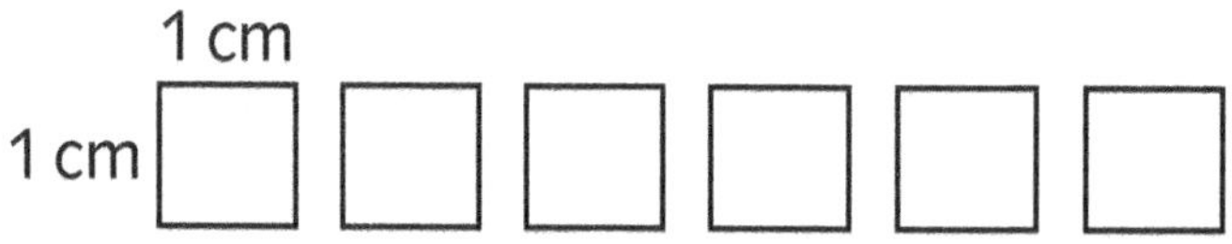

1 cm

1 cm

4 marks

Marks.......... /5

64

## Challenge 2

**PS** **1** The perimeter of the rectangular shaped Whitehill Park is 80 m.

What could the dimensions of the park be? _____

1 mark

**PS** **2** The rectangular long jump sand pit at St Lukes Primary School has a perimeter of 33 m. What could the dimensions of the sandpit be?

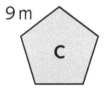

_____

1 mark

Marks............/2

## Challenge 3

**PS** **1** Look at the images below of different shaped fields.

Choose the best field, **A–C**, for each flower to be planted in.

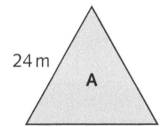

24 m

**A**

11 m

**B**

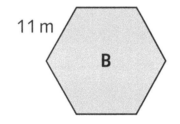

9 m

**C**

a) Bluebells need to grow in a field with a perimeter of between 60 m and 68 m.

b) Lavender needs to grow in a field with a perimeter of between 70 m and 74 m.

c) Gladioli need to grow in a field with a perimeter of less than 48 m.

3 marks

Marks........../3

Total marks ............ /10    How am I doing?

# Area

PS  Problem-solving questions

## Challenge 1

**1**  Calculate the area of these rectangles.

**a)**  8 cm

6 cm

_____

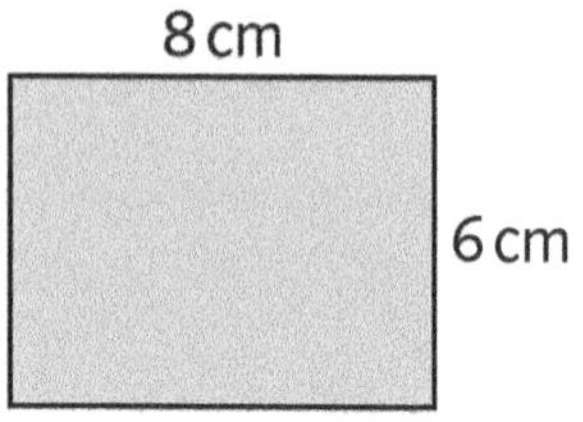

**b)**  11 cm

12 cm

_____

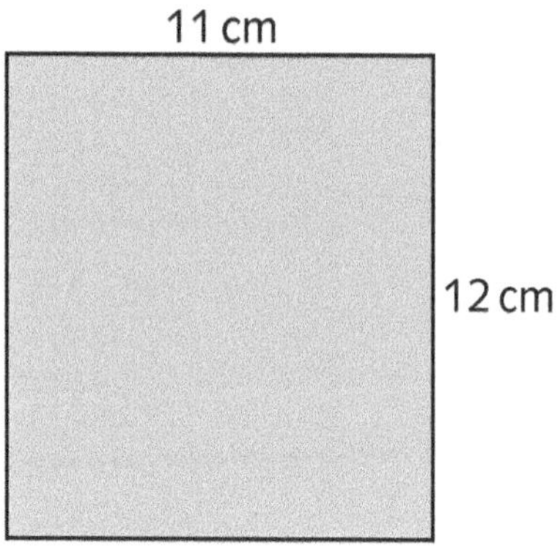

**c)**  7 cm

9 cm

_____

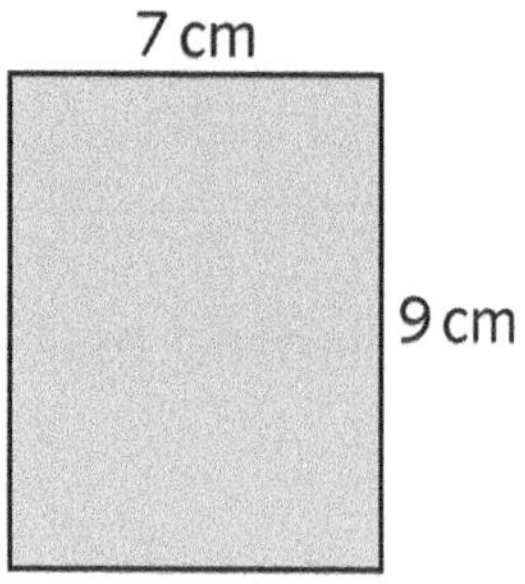

3 marks

Marks........../3

## Challenge 2

**1**  **a)**  Accurately draw a rectangle with an area of 12 cm².

**b)** Accurately draw a rectangle with an area of 18 cm².

2 marks

**PS** **2** **a)** What could the dimensions be of a rectangle with an area of 20 m²? _____

2 marks

**b)** What could the dimensions be of a rectangle with an area of 72 m²? _____

2 marks

Marks.......... /4

## Challenge 3

**PS** **1** Mrs Hendricks wants to paint the walls of the school hall. The four walls are identical in size.

One tin of paint covers 12 m². How many tins of paint will Mrs Hendricks need to cover all the walls?

← 8 m →

6 m

_____ tins

1 mark

**PS** **2** Alan is a carpet fitter. He has to fit carpet into three bedrooms. Each roll of carpet covers 8 m².

How many rolls of carpet will Alan need to carpet all the floors?

Bedroom 1: 5 m × 6 m

Bedroom 2: 9 m × 5 m

Bedroom 3: 6 m × 3.5 m

_____ rolls

1 mark

Marks.......... /2

Total marks ............ /9        How am I doing?  ☺  😐  😞

# Time

## Challenge 1

**1** Write the time shown on this clock in analogue and 24-hour form.

Analogue: _____

24-hour: _____

2 marks

**2** Write the time shown on this clock in analogue and 24-hour form.

Analogue: _____

24-hour: _____

2 marks

**3** Draw the hands on the clock showing 00:09.

1 mark

Marks.......... /5

## Challenge 2

**1** Write < or > to make these time statements correct.

a) 300 seconds ☐ $4\frac{1}{2}$ minutes

b) 450 minutes ☐ 7 hours

c) 125 hours ☐ 5 days

d) 2 days ☐ 2700 minutes

e) $\frac{1}{2}$ week ☐ 96 hours

f) 2 weeks ☐ 340 hours

6 marks

**2** Look at the data on the right. It shows the sunrise and sunset times for one week in France and Australia.

> Week of 22–28 June
> France: Sunrise at 06:32
> Sunset at 21:33
> Australia: Sunrise at 06:45
> Sunset at 16:59

a) How much more daylight did France have than Australia? _____ hours _____ minutes

b) How long was it from sunset to sunrise in Australia?

_____ hours _____ minutes

2 marks

Marks.......... /8

## Challenge 3

**1** Look at the bus timetable.

| Destination | Bus 1 | Bus 2 | Bus 3 | Bus 4 |
|---|---|---|---|---|
| Sandy Hill | 5:29 a.m. | 10:54 a.m. | 12:47 p.m. | 5:23 p.m. |
| Compton Lane | 6:39 | 11:45 | 1:58 | 6:49 |
| Buckler's Road | 7:03 | 12:36 | 2:46 | 7:23 |
| Truford Yard | 7:58 | 1:35 | 3:34 | 8:18 |
| Mulner's Way | 8:28 | 2:16 | 4:59 | 9:04 |
| Quay Point | 9:13 | 3:04 | 5:52 | 9:44 |
| Lifeboat Walk | 10:01 a.m. | 4:39 p.m. | 6:42 p.m. | 10:39 p.m. |

a) Which bus takes the shortest amount of time to get from Sandy Hill to Quay Point? _____

b) Which bus takes the longest amount of time to get from Compton Lane to Truford Yard? _____

c) If Lucy wants to be in Mulner's Way at 17:00 hours, which bus should she take? _____

d) Which is the slowest bus? _____

e) Which buses take less than $5\frac{1}{2}$ hours to get from Sandy Hill to Lifeboat Walk? _____

5 marks

Marks.......... /5

Total marks ............ /18          How am I doing?

# Time Word Problems

## Challenge 1

**PS** **1** Andy takes a flight from London to Sydney, Australia, via Hong Kong. The ticket is shown below.

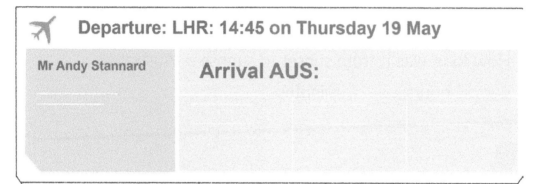

✈ **Departure: LHR: 14:45 on Thursday 19 May**

**Mr Andy Stannard**

**Arrival AUS:**

The journey takes $26\frac{1}{2}$ hours in total. What day, date and time will it be in London when Andy lands in Australia?

*3 marks*

**PS** **2** A courier has 8 hours a day to deliver as many parcels as possible. It takes 8 minutes to deliver a parcel.

How many parcels can be delivered in a day? _____

*1 mark*

Marks.........../4

## Challenge 2

**PS** **1** Maggi started work at 9:45 a.m. She worked for 8 hours. After work, she walked 5 minutes to her mum's house and stayed there for $1\frac{1}{2}$ hours. Then she walked 20 minutes back home.

What time did Maggi get home? _____

*1 mark*

# Time Word Problems

**PS** | **2** Ahmad is a nurse. He starts work at 22:25 and his night shift is $9\frac{1}{2}$ hours long. After work it takes Ahmad 15 minutes to drive home. He goes straight to sleep for 6 hours 20 minutes.

What time does Ahmad wake up? _____

1 mark

Marks............/2

## Challenge 3

**PS** | **1** If class 5SR have $2\frac{1}{2}$ hours of PE every week, how many hours do they have in 43 weeks of a school year? _____

1 mark

**PS** | **2** It takes Mira 45 minutes to complete each piece of her homework. Mira has two pieces of Literacy homework, one piece of Maths and one piece of Science homework each week. Calculate how many hours she spends on homework:

**a)** in a week [ ]

**b)** in 4 weeks [ ]

**c)** in 24 weeks [ ]

3 marks

Marks.........../4

Total marks ............ /10    How am I doing?

# Money

PS Problem-solving questions

## Challenge 1

**1** Answer these.

a) £8.83 × 100 = _____

b) £8.83 × 1000 = _____

c) £8830 ÷ 10 = _____

d) £8830 ÷ 100 = _____

4 marks

Marks.......... /4

## Challenge 2

**1** Look at the information below.

ENTRY PRICES FOR DINO WORLD:

| Adult | £9.95 | Over 60 | £5.75 |
|---|---|---|---|
| Child | £4.80 | Under 2 | £1.50 |
| Family Ticket – 2 adults + 2 children | £24.75 | Group Ticket – 5 adults, 2 over 60s, 6 children and 3 under 2s | £78.20 |

a) How much does it cost for 2 adults and 2 under 2s?

_____

b) How much does it cost for 3 adults, 2 children and 2 under 2s?

_____

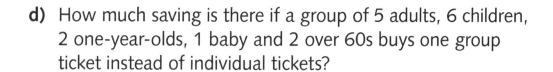

**c)** How much saving is there if a family of 2 adults and 2 children buy a family ticket?

_____

**d)** How much saving is there if a group of 5 adults, 6 children, 2 one-year-olds, 1 baby and 2 over 60s buys one group ticket instead of individual tickets?

_____

 4 marks

Marks.......... /4

## Challenge 3

**1** A packet of 10 breadsticks costs £4.10.
A box of 8 crackers costs £2.56.

**a)** How much more does 1 breadstick cost than 1 cracker?

_____

**b)** How much do 12 crackers cost? _____

2 marks

PS **2** Riley is taking part in a sponsored run. Her sponsors are going to give her 85p altogether for every lap of the school playground she runs.

**a)** How much money will Riley raise if she runs 46 laps?        _____

**b)** How many laps will Riley need to run to raise her target of £68?        _____

2 marks

Marks.......... /4

Total marks ............ /12          How am I doing?

# Volume

### Challenge 1

**1** Estimate the volume of these cuboids.

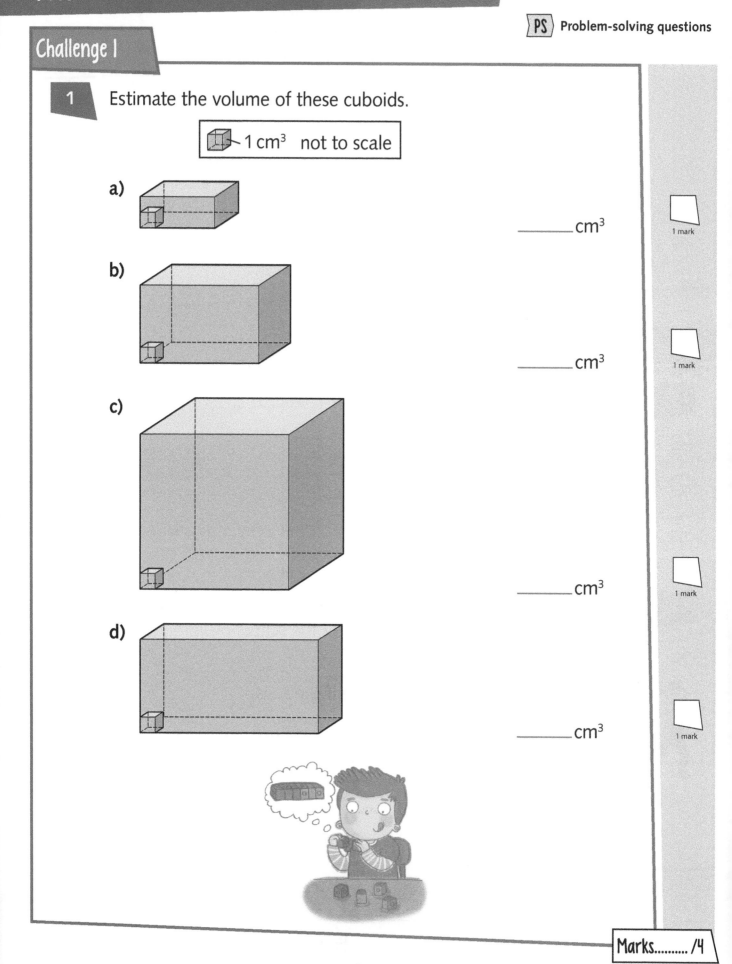

1 cm³  not to scale

a)

_____ cm³

1 mark

b)

_____ cm³

1 mark

c)

_____ cm³

1 mark

d)

_____ cm³

1 mark

Marks.........../4

74

# Volume

## Challenge 2

**1** Find the volume of these solids.

 — 1 cm³ not to scale

**a)**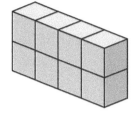

_____ cm³

**b)**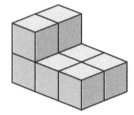

_____ cm³

**c)**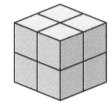

_____ cm³

**d)**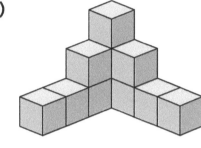

_____ cm³

4 marks

Marks.......... /4

## Challenge 3

**PS** **1** Find one possible set of dimensions for a cuboid with the following volume:

**a)** 20 cm³ _____

**b)** 30 cm³ _____

**c)** 100 cm³ _____

3 marks

Marks.......... /3

Total marks ............. /11     How am I doing?

# Equivalent and Imperial Units of Measure

## Challenge 1

**1** Write the imperial units under the correct headings to show what each measures.

| pints | yards | stones | feet | ounces | miles |
| pounds | fluid ounces | inches | gallons |

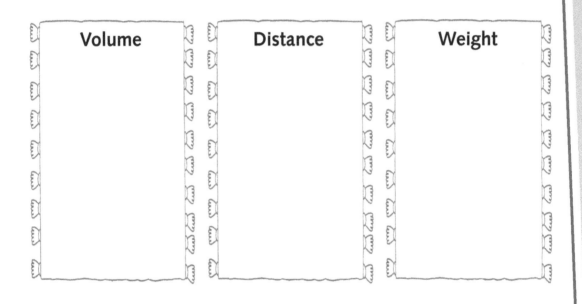

| Volume | Distance | Weight |
|--------|----------|--------|
|        |          |        |

10 marks

Marks......... /10

## Challenge 2

**1** How many inches are there in the following measures?

**a)** 6 feet ☐

**b)** 19 feet ☐

**c)** 22 feet ☐

**d)** 128 feet ☐

**e)** 2404 feet ☐

There are 12 inches in one foot.

5 marks

# Equivalent and Imperial Units of Measure

**2** How many inches are there in the following measures?

a) 6 yards [ ]

b) 40 yards [ ]

c) 93 yards [ ]

d) 349 yards [ ]

e) 5820 yards [ ]

There are 36 inches in one yard.

5 marks

Marks.........../10

## Challenge 3

**1** Convert the measurements to complete the table.

|  | cm | inches (approx.) |
|---|---|---|
| Height | 127 cm |  |
| Foot | 30 cm |  |
| Hand span | 18 cm |  |

2.5 cm = approx. 1 inch

3 marks

Marks.........../3

Total marks ............./23

How am I doing?

## Measurement Word Problems

### Challenge 1

**PS** **1** Tyson is planning to run a marathon, which is 26 miles. He has found 9 people who are each going to sponsor him £3 for every mile he runs. How much money will he raise if he completes the marathon?

£_____

*1 mark*

**PS** **2** Mr Hawkins wants all of the five office floors laminated. Each floor measures 8 m × 6 m. One pack of laminate will cover 12 m². How many packs of laminate will Mr Hawkins need to cover these floors?

a) One room [ ]

b) Five rooms [ ]

*2 marks*

Marks........../3

### Challenge 2

**PS** **1** Dympna needs 8 pounds of flour and 3 pounds of sugar for a recipe. If there are 16 ounces in a pound, how many ounces of flour and sugar does Dympna need altogether?

_____

*1 mark*

**PS** **2** The council want to paint the outline of a field with white paint. The area of the rectangular field is 192 m².

16 m

?

Each tin of white paint will only cover 8 m. How many tins of paint are needed to paint the perimeter of the field? _____

*2 marks*

Marks........../3

# Measurement Word Problems

**Challenge 3**

PS **1** A woman wants to put up a new fence at the sides and at the bottom of her garden.

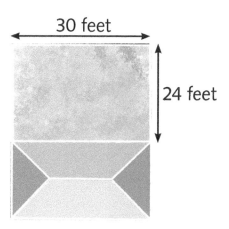

30 feet

24 feet

A fence panel measures 6 ft wide. How many fence panels are needed to secure the three sides of this garden?

_____

1 mark

PS **2** At the warehouse, Barry is wrapping two boxes of furniture to send to a customer.

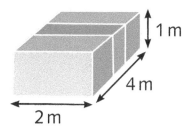 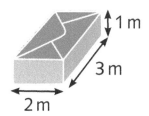

1 m
4 m
2 m

1 m
3 m
2 m

Each roll of wrapping paper will cover 5 m² of the boxes. How many rolls of wrapping paper will Barry need to cover both boxes of furniture?

_____ rolls

3 marks

Marks.......... /4

Total marks ............. /10          How am I doing?   😊  😐  😣

# Progress Test 3

1. How many ounces are there in $\frac{1}{4}$ of a pound? _____

   *1 mark*

2. A playground measures 21 m × 17 m. Calculate the following:

   a) Area [ ]

   b) Perimeter [ ]

   *2 marks*

3. A Maths lesson is an hour long and begins at 10:20 a.m. What time does the fire bell ring if it goes off $\frac{2}{3}$ of the way through the lesson? _____

   *1 mark*

4. Estimate the volume of this cuboid. _____

   ▧ 1 cm³

   *1 mark*

5. A rectangular swimming pool has an area of 108 m². What could the dimensions be?

   _____

   *1 mark*

6. If the time on an analogue clock is 8:26 p.m. what is the time on a 24-hour clock? _____

   *1 mark*

7. If it costs £12.80 for 4 children to go to the cinema, how much does it cost for 1 child?     £_____

   *1 mark*

80

**PS** **8.** Darius wants to paint one of the walls of his office. The wall measures 9 m × 5.5 m. If 1 tin of paint will cover 4 m², how many tins of paint will Darius need? _____

1 mark

**PS** **9.** Esmae is organising a charity bake sale and hopes to raise £86. She will sell small cakes for 80p each and large cakes for £2.50. If Esmae sells 28 large cakes, how many small cakes will she need to sell in order to reach her target?

_____

1 mark

**10.** How many inches are there in 32 yards? _____

1 mark

**11.** Draw a plan of a rectangular school hall that has a perimeter of 42 m.

1 mark

**12.** Calculate 362 × 28.

_____

1 mark

**13.** Write this year in numerals.

MDXCIV

_____

1 mark

**14.** Write < or > in the box to make the statement correct.

$\frac{5}{8}$ ☐ $\frac{3}{4}$

1 mark

**15.** Write < or > in the box to make the statement correct.

15 weeks ☐ 107 days

1 mark

**16.** How many 45 minute sessions of piano practice will Dawn need to do if her teacher says she must do at least 19 hours altogether? _____

1 mark

**17.** Arrange these five 4 cm squares so that they have a perimeter of 48 cm.

Not to scale   4 cm ☐ ☐ ☐ ☐ ☐

4 cm

1 mark

**18.** Draw the hands on the clock to show 2 hours 43 minutes later than 19:57.

1 mark

**19.** Write $\frac{29}{6}$ as a mixed number.

1 mark

**20.** Calculate 3421 × 26.

_____

1 mark

**PS** **21.** If it costs six adults £102 to travel by train to London, how much does one adult ticket cost?     £_____

1 mark

**22.** Write this year in numerals.

MMXXI

_____

1 mark

**23.** How many inches are there in 28 feet?     _____ inches

1 mark

**PS** **24.** Class 5LH are taking part in a sponsored silence. Each pupil is sponsored 26p for every hour they remain silent. How much money will the class raise if all 32 pupils remain silent for $4\frac{1}{2}$ hours?     £_____

1 mark

**25.** Calculate 1978 × 35.

_____

1 mark

**26.** What would the time be on a clock if it was 5 hours 12 minutes earlier than 00:32?

_____

1 mark

**27.** Calculate $\frac{2}{5}$ × 6.

1 mark

Marks........ /28

# Regular and Irregular Polygons

## Challenge 1

**1** Draw a line from each shape to the correct word to show whether the polygon is regular or irregular.

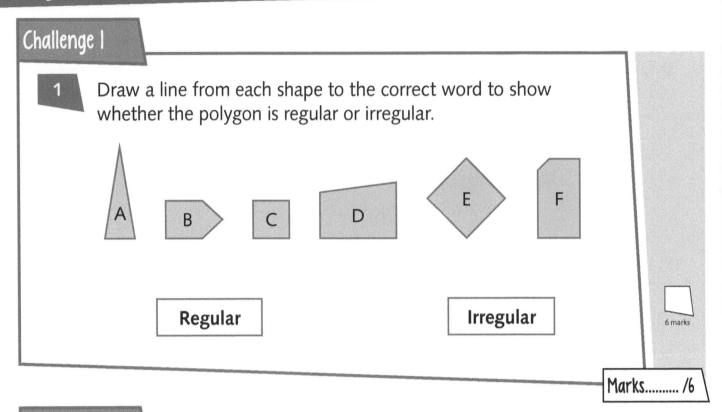

Regular                    Irregular

6 marks

Marks.......... /6

## Challenge 2

**1** Sketch an example of each polygon as it is in its regular and its irregular form. Angle sizes do not need to be exact.

| Polygon | Regular | Irregular |
|---|---|---|
| Triangle | | |
| Quadrilateral | | |
| Pentagon | | |

# Regular and Irregular Polygons

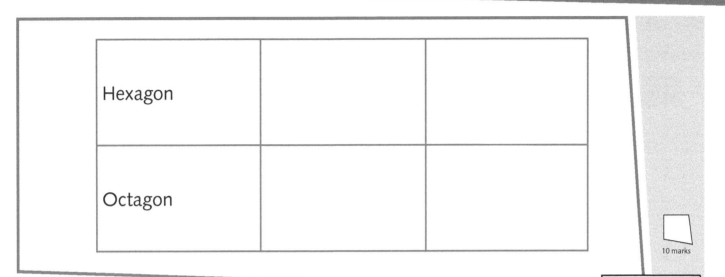

| Hexagon | | |
| --- | --- | --- |
| Octagon | | |

10 marks

Marks......... /10

**1** Calculate the perimeter of this irregular pentagon.

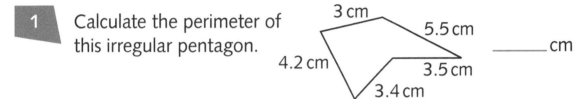

3 cm
5.5 cm
4.2 cm
3.5 cm
3.4 cm

_____ cm

1 mark

**2** Calculate the perimeter of this irregular hexagon.

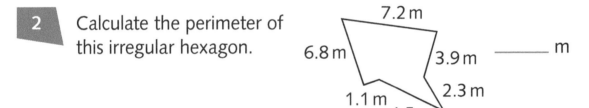

7.2 m
6.8 m
3.9 m
1.1 m
2.3 m
4.5 m

_____ m

1 mark

**3** This irregular octagon has a perimeter of 49 cm. Calculate the length of the missing side.

_____ cm

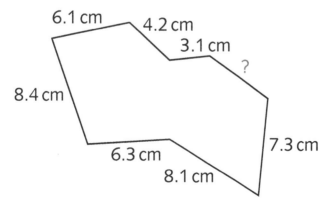

6.1 cm
4.2 cm
3.1 cm
?
8.4 cm
7.3 cm
6.3 cm
8.1 cm

1 mark

Marks......... /3

Total marks ............ /19

How am I doing?

# 2-D Shapes

## Challenge 1

 **1** Some shapes and clues are given below. Write each 2-D shape into the correct place in the grid using the clues provided. Include a drawing and the name of the shape.

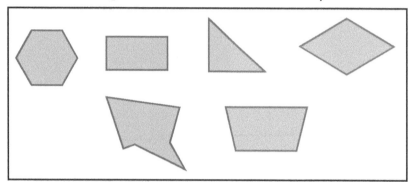

**Clue 1:** The trapezium is in the top right-hand box.

**Clue 2:** The shape with 4 equal angles and 2 pairs of sides of equal length is in box number 5.

**Clue 3:** The shape with all of its angles of equal size and sides of equal length is in box number 4.

**Clue 4:** The shape with a right angle but no parallel sides is in the box above the rectangle.

**Clue 5:** The irregular hexagon is in the box under its regular version.

**Clue 6:** The shape that has matching opposite acute and obtuse angles is in box number 1.

| 1. | 2. |
|---|---|
| 3. | 4. |
| 5. | 6. |

6 marks

Marks.......... /6

# 2-D Shapes

## Challenge 2

**PS** **1** How many isosceles triangles are needed to make an octagon? _____

1 mark

**PS** **2** How many isosceles triangles are needed to make a pentagon? _____

1 mark

**PS** **3** What is the smallest number of triangles needed to make a rhombus? _____

1 mark

Marks.........../3

## Challenge 3

**1** Look at the shapes below.

hexagon    rectangle    triangle    rhombus    irregular hexagon

parallelogram    circle    pentagon    octagon    trapezium

Which shape has the most of the following?

**a)** Lines of symmetry _____

**b)** Right angles _____

**c)** Pairs of parallel lines _____

**d)** Vertices _____

4 marks

Marks.........../4

Total marks ............ /13          How am I doing?  😊  😐  😫

# 3-D Shapes

## Challenge 1

**1** Draw lines to match each property to the correct 3-D shape.

**A:** It has identical triangular faces at each end, 5 faces and 6 vertices

**B:** It has a triangular base, 4 vertices and 4 faces

**C:** It has 2 identical circular faces and no vertices

**D:** It has 6 identical square faces and 8 vertices

**E:** It has a circular base, 1 apex and 1 curved surface

**F:** It has a square base, 5 vertices and 5 faces

**G:** It has 1 curved surface, no edges

**1:** Tetrahedron

**2:** Cube

**3:** Triangular prism

**4:** Square-based pyramid

**5:** Cone

**6:** Sphere

**7:** Cylinder

7 marks

Marks............/7

## Challenge 2

**1** Which 3-D shapes are these?

a)  _____

b)  _____

c)

d)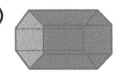

_____          _____

4 marks

Marks............/4

# 3-D Shapes

### Challenge 3

**1** Complete the table for a hexagonal prism.

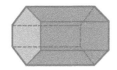

| Number of edges | Number of faces | Number of vertices |
|---|---|---|
|  |  |  |

**2** Complete the table for an octahedron.

| Number of edges | Number of faces | Number of vertices |
|---|---|---|
|  |  |  |

3 marks

**3** Complete the table for a pentagonal prism.

| Number of edges | Number of faces | Number of vertices |
|---|---|---|
|  |  |  |

 3 marks

Marks.......... /9

Total marks ............. /20       How am I doing?   😊   😐   😣

# Properties of Rectangles

## Challenge 1

**1**  Circle **True** or **False** for each statement.

a) Rectangles are 3-D shapes.                                      **True / False**

b) A rectangle is an oblong.                                       **True / False**

c) A rectangle is a quadrilateral.                                 **True / False**

d) A rectangle is an irregular shape.                              **True / False**

e) A rectangle has four equal sides.                              **True / False**

f) Rectangles have four equal angles.                            **True / False**

g) The angles in a rectangle add up to 380°.                     **True / False**

h) A rectangle has one pair of parallel sides.                   **True / False**

i) The sides on either side of a right angle
are perpendicular to one another.                              **True / False**

9 marks

Marks.......... /9

## Challenge 2

**1**  If the area of this rectangle is 30 m², what
could the lengths of the sides be?

_____

1 mark

**2**  If the perimeter of this rectangle is 28 cm, what
could the lengths of the sides be?

_____

1 mark

Marks.......... /2

# Properties of Rectangles

**Challenge 3**

**1** Look at the compound shape below.

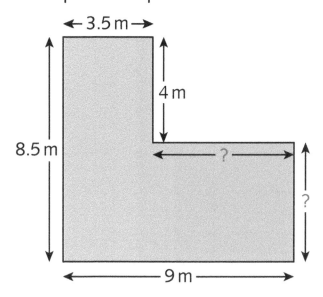

a) Work out the missing lengths. _____

b) Calculate the perimeter. _____ m

c) Calculate the area. _____ m²

**2** Calculate the area and perimeter of this compound shape.

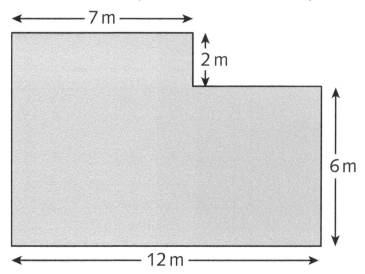

Area = _____ m²          Perimeter = _____ m

Marks.......... /6

Total marks ............. /17          How am I doing?

# Different Types of Angle

**1** Name each angle and estimate its size.

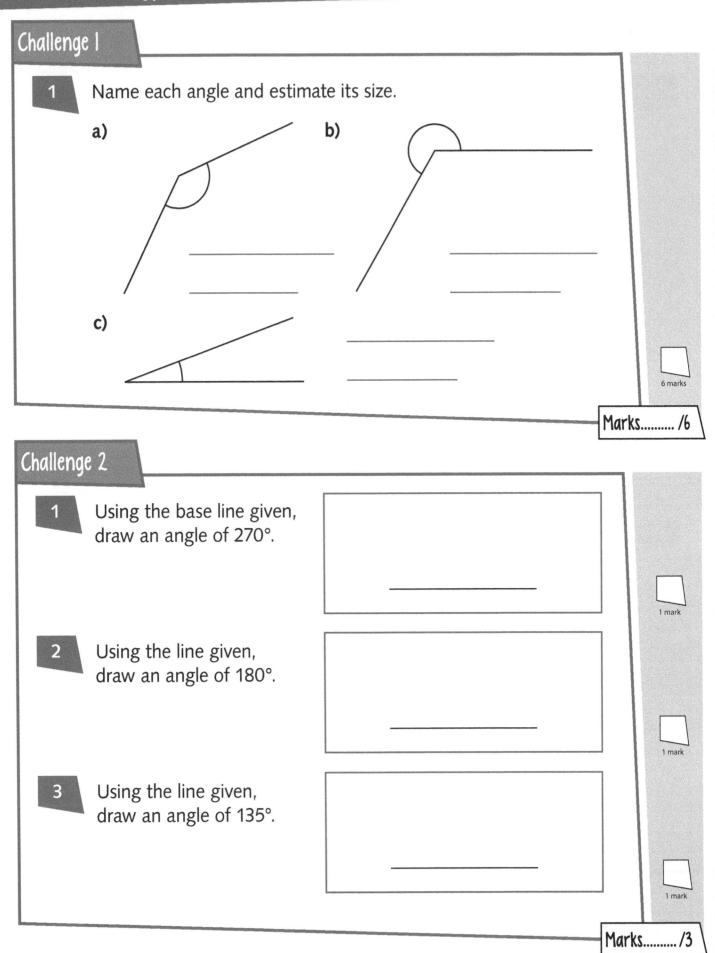

a)

_____

_____

b)

_____

_____

c)

_____

_____

6 marks

Marks.......... /6

**Challenge 2**

**1** Using the base line given, draw an angle of 270°.

_____

1 mark

**2** Using the line given, draw an angle of 180°.

_____

1 mark

**3** Using the line given, draw an angle of 135°.

_____

1 mark

Marks.......... /3

# Different Types of Angle

**1** Draw an irregular pentagon that has 1 right angle, 2 acute angles, 1 obtuse angle and 1 reflex angle.

2 marks

**2** Draw an irregular hexagon that has 1 right angle, 3 obtuse angles, 1 reflex angle and 1 acute angle.

2 marks

Marks......... /4

Total marks ............ /13          How am I doing?

# Calculating Angles

## Challenge 1

**1** How many degrees does the minute hand rotate through in 60 minutes?

_____°

**2** What is the angle at the centre of each slice if a pizza is cut into fifths?

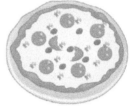

_____°

**3** Between the times 01:30 and 04:00, how many degrees has the minute hand on a clock moved through?

_____°

Marks.........../3

## Challenge 2

**1** Look at this table.

| Shape | Total degrees |
|-----------|---------------|
| Triangle | 180° |
| Rectangle | 360° |
| Pentagon | 540° |
| Hexagon | 720° |

Use the table to help you calculate how many degrees are in each angle in these shapes.

**a)** Equilateral triangle _____°

**b)** Rectangle _____°

**c)** Regular pentagon _____°

**d)** Regular hexagon _____°

Marks.........../4

# Calculating Angles

**1** Angles around a point total 360°. Calculate the missing angle in each of these.

**a)**

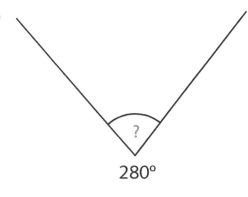

280°

_____°

**b)**

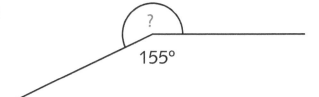

155°

_____°

**c)**

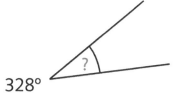

328°

_____°

3 marks

Marks............/3

Total marks ............. /10          How am I doing?

# Co-ordinates

**1** **a)** What is the co-ordinate of the point midway between (2, 4) and (2, 10)? _____

**b)** What is the co-ordinate of the point midway between (4, 5) and (10, 11)? _____

 2 marks

**2** **a)** Plot these co-ordinates on the grid below.

**(3, 5)   (4, 9)   (5, 5)   (6, 9)**

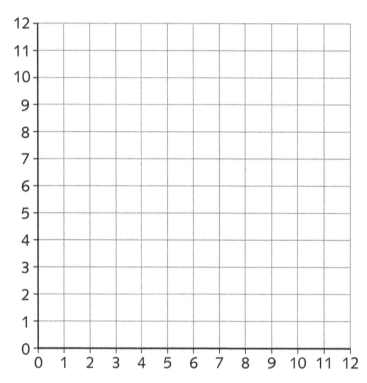

**b)** What shape do they make? _____

1 mark

1 mark

Marks.......... /4

**1** Fill in the missing co-ordinates so that each set will make a rectangle.

**a)** (2, 2) (7, 4) (7, 2) (_____, _____)

**b)** (6, 5) (6, 1) (4, 1) (_____, _____)

2 marks

# Co-ordinates

**2** Some treasure is hidden at the point of the missing vertex of this square: (1, 5) (4, 5) (1, 8). What is the co-ordinate of the treasure?

(_____, _____)

1 mark

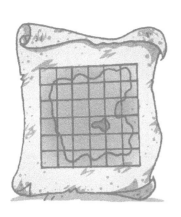

Marks............/3

## Challenge 3

**1** Write the co-ordinate that is needed to make each set into a rhombus.

a) (10, 9) (5, 6) (6, 10) (_____, _____)

b) (5, 4) (3, 8) (5, 12) (_____, _____)

c) (12, 18) (15, 16) (12, 14) (_____, _____)

3 marks

**2** What is the perimeter of a rectangle positioned at (7, 6) (3, 6) (3, 8) (7, 8) on a 1 cm² grid?

_____ cm

1 mark

Marks............/4

Total marks ............ /11          How am I doing?

# Translations

## Challenge 1

**1** Look at the grid below.

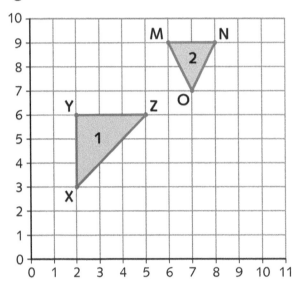

a) Shape 1 is translated. If point **X** is translated from (2, 3) to (6, 5), what are the new co-ordinates of the other two points?

Y (_____ , _____)  Z (_____ , _____)

b) Shape 2 is translated. If point **O** is translated from (7, 7) to (4, 5), what are the new co-ordinates of the other two points?

M (_____ , _____)  N (_____ , _____)

4 marks

Marks.......... /4

## Challenge 2

**1** Look at the grid below.

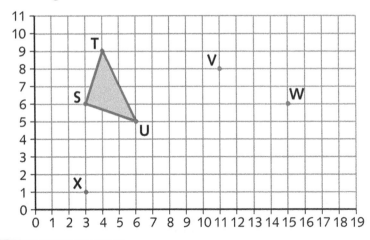

# Translations

a) If point **S** is translated to point **V**, what will the new co-ordinates be of the other two points?

**T** (_____, _____)    **U** (_____, _____)

b) If point **T** is translated to point **W**, what will the new co-ordinates be of the other two points?

**S** (_____, _____)    **U** (_____, _____)

c) If point **U** is translated to point **X**, what will the new co-ordinates be of the other two points?

**S** (_____, _____)    **T** (_____, _____)

6 marks

Marks.......... /6

## Challenge 3

**1** If the co-ordinates of a shape after the translation up 4, right 8 are (11, 10) (12, 13) (14, 10) (9, 13), what was the original position of the shape?

(_____, _____)    (_____, _____)

(_____, _____)    (_____, _____)

4 marks

**2** If the co-ordinates of a shape after the translation left 6, down 6 are (2, 5) (5, 7) (6, 5), what was the original position of the shape?

(_____, _____)    (_____, _____) (_____, _____)

3 marks

**3** If the co-ordinates of a shape after the translation up 2, right 6 are (7, 4) (9, 4) (9, 7) (7, 7), what was the original position of the shape?

(_____, _____)    (_____, _____)

(_____, _____)    (_____, _____)

4 marks

Marks.......... /11

Total marks ............ /21          How am I doing?

# Reflection

## Challenge 1

**1** Look at the grid.

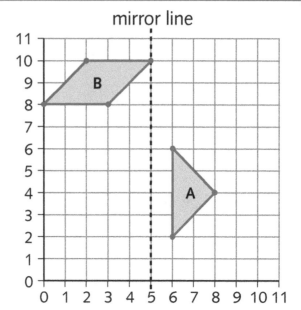

mirror line

**a)** What are the new co-ordinates of shape A after reflection in the mirror line?

(_____ , _____)

(_____ , _____)

(_____ , _____)

3 marks

**b)** What are the new co-ordinates of shape B after reflection in the mirror line?

(_____ , _____) (_____ , _____) (_____ , _____) (_____ , _____)

4 marks

Marks............/7

## Challenge 2

**1** Look at this grid showing a parallelogram.

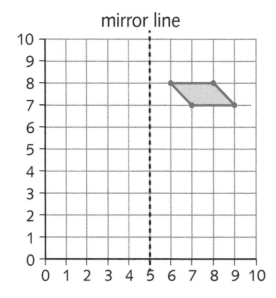

mirror line

# Reflection

Write the new co-ordinates of the parallelogram:

**a)** after reflection in the mirror line.

(_____ , _____) (_____ , _____) (_____ , _____) (_____ , _____)

**4 marks**

**b)** after reflection and then translation down 5, right 3.

(_____ , _____) (_____ , _____) (_____ , _____) (_____ , _____)

**4 marks**

Marks.........../8

## Challenge 3

**1** Use this grid for parts **a)** and **b)**.

mirror line

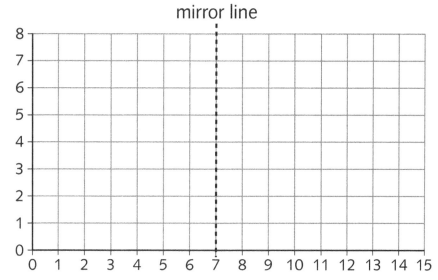

**a)** Plot these co-ordinates. They are three vertices of a rhombus.

(7, 2) (10, 3) (11, 6) (_____ , _____)

Write the missing co-ordinate to complete the rhombus.

**2 marks**

**b)** What are the co-ordinates of the rhombus after it is reflected in the mirror line?

(_____ , _____) (_____ , _____) (_____ , _____) (_____ , _____)

**4 marks**

Marks.........../6

Total marks ............ /21     How am I doing?

# Tables and Timetables

## Challenge 1

**1**  Below is a table showing the number of house points that each house in St Henry's School have been awarded each year.

| Year | Red House | Blue House | Yellow House | Green House |
|------|-----------|------------|--------------|-------------|
| 2010 | 89 | 96 | 103 | 101 |
| 2011 | 102 | 100 | 110 | 106 |
| 2012 | 106 | 101 | 115 | 122 |
| 2013 | 120 | 121 | 98 | 98 |
| 2014 | 119 | 119 | 119 | 99 |
| 2015 | 101 | 101 | 89 | 95 |
| 2016 | 92 | 107 | 113 | 124 |

a)  Which house has performed the best overall?  _____

b)  How many points separate the highest and lowest performing houses?  _____

c)  In which year was the total for all four houses altogether the highest?  _____

3 marks

Marks.........../3

## Challenge 2

**1**  These are the results from a world athletics championships. There are medals for the first three places in each event.

| Event | India | Britain | America | Ethiopia | Spain | Japan |
|-------|-------|---------|---------|----------|-------|-------|
| 100 m sprint (sec) | 13.8 | 14.01 | 14.2 | 12.98 | 12.84 | 13.5 |
| 1500 m long distance (min) | 5.32 | 6.53 | 7.23 | 6.2 | 5.49 | 5.55 |
| Shot put (metres) | 5.12 | 5.92 | 6.45 | 8.10 | 7.65 | 7.35 |
| High jump (metres) | 1.36 | 0.95 | 0.86 | 1.04 | 1.53 | 1.12 |
| Triple jump (metres) | 7.3 | 8.93 | 7.88 | 8.17 | 7.75 | 7.9 |

# Tables and Timetables

a)  Which country won the 100 m sprint?  _____

b)  Which country didn't win any medals?  _____

c)  How much further did the Ethiopian
shot-putter throw than the British athlete?  _____ m

3 marks

Marks.......... /3

## Challenge 3

  a)  Complete this frequency table showing how many pages
children in Years 5 and 6 read in a set time.

| Class | Number of pages read | Total pages read |
|-------|---------------------|------------------|
| 5HY | | 68 |
| 5AS | 卌 卌 卌 卌 卌 卌 卌 卌<br>卌 卌 卌 卌 卌 卌 卌 卌 | | 
| 5LS | | 86 |
| 5LM | 卌 卌 卌 卌 卌 卌 卌<br>卌 卌 卌 III | |
| 6JS | 卌 卌 卌 卌 卌 卌 卌<br>卌 卌 卌 卌 卌 卌 卌 IIII | |
| 6PD | | 83 |
| 6KF | | 79 |
| 6SH | 卌 卌 卌 卌 卌 卌 卌 卌<br>卌 卌 卌 卌 卌 卌 卌 卌 II | |

b)  Which class read the most pages in the time?  _____

c)  What is the total amount of pages read in both
year groups altogether?  _____

d)  How many more pages did Year 6 read than Year 5?  _____

8 marks

1 mark

1 mark

1 mark

Marks.......... /11

Total marks ............ /17     How am I doing?

# Bar Charts

Look at this bar chart and use it to answer the questions in Challenges 1–3.

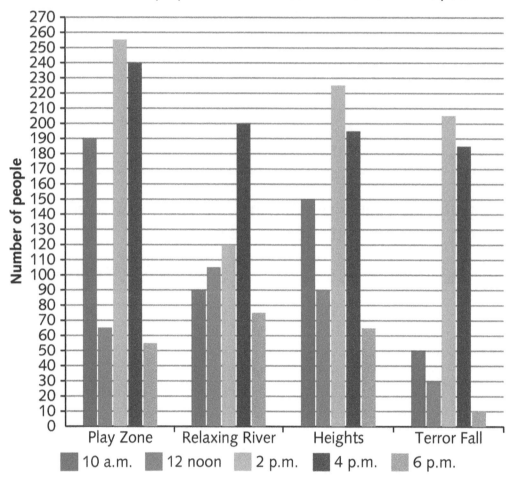

The number of people that visited different zones in a theme park

■ 10 a.m.   ■ 12 noon   ■ 2 p.m.   ■ 4 p.m.   ■ 6 p.m.

## Challenge 1

**1**   Complete this table showing the data in the bar chart.

| Zone | Number of people at different times of day | | | | |
|------|---------|-----------|---------|---------|---------|
|      | 10 a.m. | 12 noon | 2 p.m. | 4 p.m. | 6 p.m. |
| **Play Zone** | 190 | | | | |
| **Relaxing River** | | | | | |
| **Heights** | | | | | |
| **Terror Fall** | | | | | 10 |

18 marks

Mark .......... /18

# Bar Graphs

**1** **a)** Which zone was the busiest throughout the day?

_____

**b)** Which zone saw the biggest difference in visitors throughout the day?

_____

**c)** How many more people visited the busiest zone compared to the least busy zone?

_____ people

3 marks

Marks............/3

**1** **a)** At what time were all the zones the least busy? _____

**b)** Suggest a reason why that might be.

_____

2 marks

**2** **a)** Which zone was the least popular throughout the day?

_____

**b)** Suggest a reason why this zone may have seen the smallest number of visitors.

_____

2 marks

Marks........../4

Total marks ............ /25     How am I doing?

# Line Graphs

**1** Draw a line to match each line graph to the correct statement.

**A**

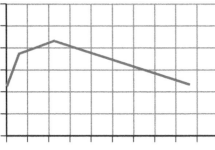

**1**

I started off steadily then I went into a sprint and jog rhythm, then I took a steady pace home.

**B**

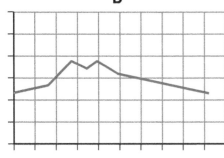

**2**

I ran quickly, then walked at a constant pace, then went into another quick run, then ran at different slower paces all the way home.

**C**

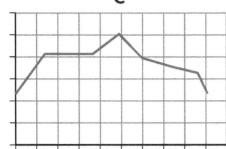

**3**

I ran at three different speeds.

3 marks

Marks........../3

**1** Draw a line graph to represent the journey of a car that sets off in a residential area, then drives along the motorway but then gets caught in a traffic jam. When the jam clears, the car continues on the motorway.

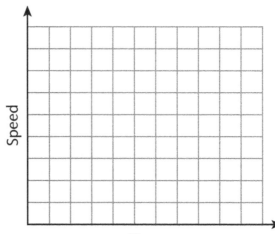

4 marks

Marks........../4

# Line Graphs

**Challenge 3**

**1**

**a)** Use the data in the table to plot a line graph showing the distance that a marathon cyclist covered during a race.

| Distance (miles) | 8 | 16 | 23 | 29 | 33 | 35 | 37 | 38 | 40 | 43 |
|---|---|---|---|---|---|---|---|---|---|---|
| Time (hours) | 0.5 | 1 | 1.5 | 2 | 2.5 | 3 | 3.5 | 4 | 4.5 | 5 |

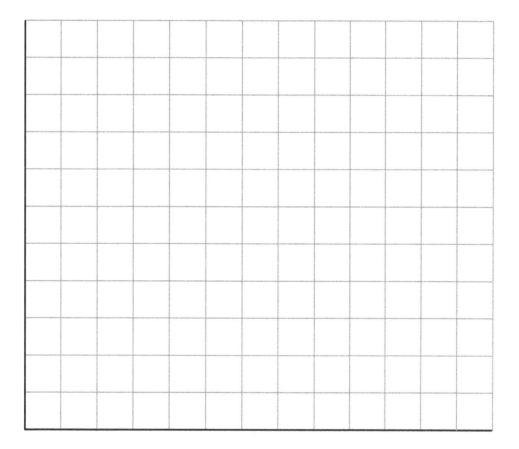

4 marks

**b)** Use your graph to predict the distance the cyclist had covered after $6\frac{1}{2}$ hours. _____

1 mark

Marks.......... /5

Total marks ............. /12

How am I doing?

1. Round 36.64 to the nearest whole number. _____

   1 mark

2. Draw an obtuse angle.

   1 mark

3. Look at this co-ordinate grid.

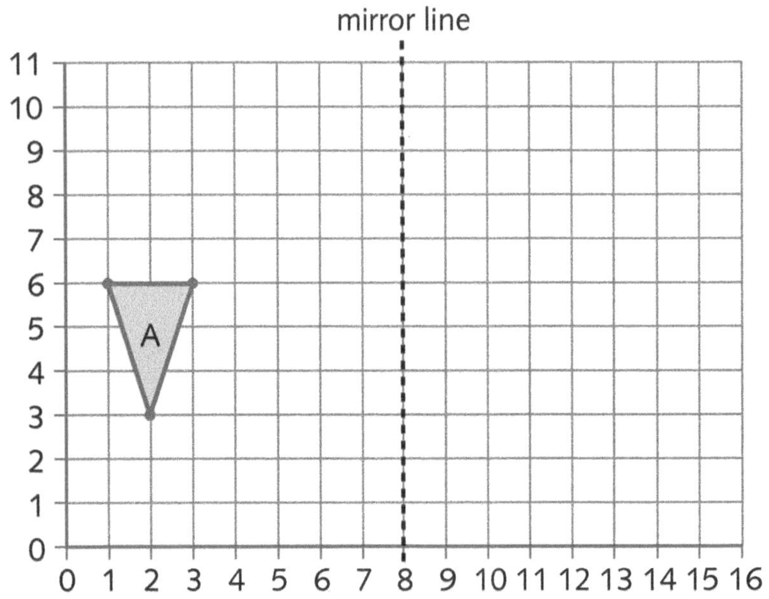

   mirror line

   a) What are the new co-ordinates of triangle A after translation up 4, right 8?

      (_____, _____) (_____, _____) (_____, _____)

      3 marks

   b) What are the co-ordinates of triangle A after reflection in the mirror line?

      (_____, _____) (_____, _____) (_____, _____)

      3 marks

4. Draw an irregular polygon with 2 acute angles and 2 obtuse angles.

   1 mark

**5.** What are the measurements in degrees that make a reflex angle?

_____

1 mark

**6.** What 2-D shapes are needed to make
the net of a cylinder?

_____

1 mark

**7.** What are the missing co-ordinates needed to make a rectangle?

(9, 4)     (11, 4)     (11, 7)     (_____, _____)

1 mark

**8.** A local community group are running an event to raise money for
the youth club. These are the total amounts raised from each stall.

| Cake Sale | £49.59 | Tombola | £120 |
|---|---|---|---|
| Nearly New | £52.95 | Teas and Coffees | £83.12 |
| Splat the Rat | £11.02 | Handmade Cards | £39.19 |
| Treasure Island | £28.49 | Kids' Play Area | £20 |
| How many marbles in the jar? | £12.94 | Candy Buffet | £19.47 |
| Guess the name of the teddy | £13.50 | Make your own jewellery | £23.84 |

**a)** How much money was raised altogether? £_____

**b)** How much more money did the Tombola and the Cake Sale
make than the Handmade Cards?

£_____

**c)** It was hoped that the Nearly New stall would raise £35. How
much over the target did they raise to the nearest pound?

£_____

3 marks

# Progress Test 4

PS Problem-solving questions

**9.** Draw a labelled line graph to represent the journey of this aeroplane:

After taking off, the plane travelled at a steady pace, it then travelled faster for a while before returning to a steady pace and slowly coming in to land.

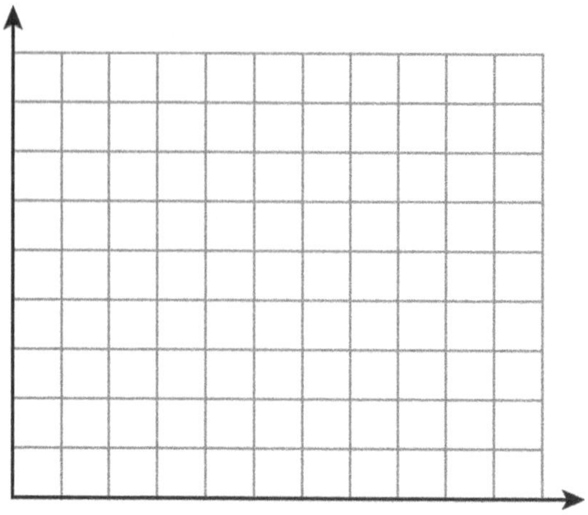

1 mark

**10.** Using the line given, draw an angle of 45°.

1 mark

**11.** Draw an irregular pentagon that has at least one angle of every type (acute, obtuse, right, reflex).

4 marks

**12.** If a cake is cut into eighths, what is the angle at the centre of each piece?

______°

1 mark

**13.** What is the angle between the hands of a clock at 1 o'clock?

______°

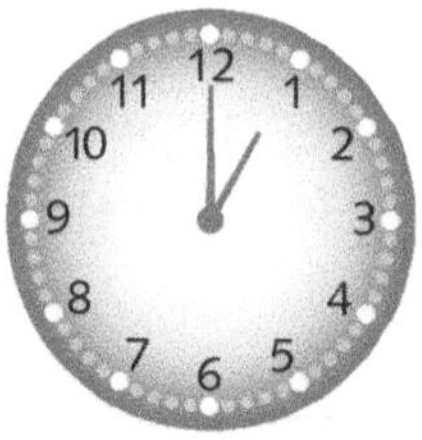

1 mark

**14.** Write the decimal and fraction equivalents for 25%. _____ _____

**15.** If it takes Callum 34.5 seconds to swim 50 m, how long will it take him to swim 200 m if he swims at a constant speed?

_____

**16.** Which 2-D shapes make up the faces of a hexagonal prism? How many are there of each?

_____

**17.** Calculate 4273 × 25.

_____

**18.** How many edges does a triangular-based prism have? _____

**19.** Look at this diagram of a field.

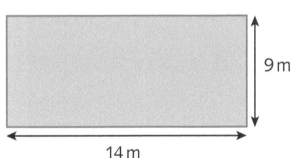

9 m

14 m

    **a)** Calculate the area. _____ m²

    **b)** Calculate the perimeter. _____ m

**20.** Solve $\frac{8}{12} \times 4$.

_____

**21.** What is the perimeter of this irregular pentagon?

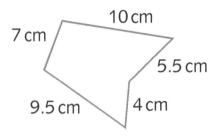

10 cm

7 cm

5.5 cm

9.5 cm　　4 cm

_____ cm

# Notes

# Answers

**Pages 4–11**
**Starter Test**
1. 10.42, 10.39, 10.24, 10.06, 10.00
2. 3459
3. 4
4. 4
5. 9 minutes
6. 400 ml
7. $\frac{1}{8}$, $\frac{3}{8}$, $\frac{4}{8}$, $\frac{5}{8}$
8. 4 × 2.5 = 10
9. 8500
10. 8 weeks
11. 5
12. Train 2
13. Any 12 parts shaded, e.g.

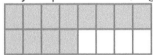

14. 149 278 − 1000 = 148 278 and
    148 278 − 1000 = 147 278
15. 24 boxes
16. £4.76
17. Any three appropriate fractions, e.g. $\frac{8}{10}$, $\frac{9}{10}$, $\frac{95}{100}$
18. 26 m
19. £2.00
20. 84
21. 5
22. a) ÷ 100
    b) ÷ 10
23. Possible answers:

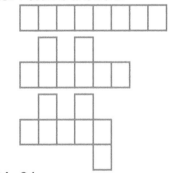

24. 84
25. 330 minutes
26. 3 hours
27. 11:05 a.m.
28. 6
29. e.g. $\frac{4}{8}$
30. a) 15
    b) 39
31. 7638
32. 15 rows
33. Possible answers include: 12 m × 26 m,
    19 m × 19 m, 24 m × 14 m

34. 960
35. 72
36. $\frac{1}{3}$ or $\frac{6}{18}$
37. 301
38. £30
39. Twenty-six thousand three hundred and
    ninety-seven
40. 12:20 p.m.

**Pages 12–13**
**Challenge 1**
1. Fourteen thousand two hundred and eighty
   seven
2. 37.92, 379.3, 3710, 3789, 3792
3. 28.5, 42.3
**Challenge 2**
1. a) 70
   b) 5 hundredths
2. 1 392 482, 37 476, 583 406.87 circled
3. Lilly's score
**Challenge 3**
1. a) 79 497
   b) 79 506
   c) 79 596
   d) 80 496
   e) 89 496
   f) 179 496
2. a) 4895
   b) 489.5
   c) 48.95
   d) 4.895
3. a) <
   b) <
   c) >
   d) <
4. 783

**Pages 14–15**
**Challenge 1**
1. a) 29 changes to 30
   b) 3729 changes to 4000
2. a) 7000
   b) 27 000
   c) 19 000
   d) 329 000
   e) 743 000
   f) 38 000
3. a) 80 000
   b) 30 000
   c) 190 000
   d) 290 000
   e) 280 000
   f) 560 000

# Answers

4. a) 300000
   b) 300000
   c) 100000
   d) 500000
   e) 900000

**Challenge 2**
1. a) 34999
   b) 25000
2. a) 114.7
   b) 93.4

**Challenge 3**
1. a) £16
   b) £21
   c) £4
   d) £1287
   e) £2375
   f) £12630
2. a) 3
   b) 30
   c) 37
   d) 694
   e) 947
   f) 47695
3. a) 132.9
   b) 1237.8
4. 74000
5. 180000

**Pages 16–17**
**Challenge 1**
1. 1 – I
   5 – V
   10 – X
   15 – XV
   20 – XX
   28 – XXVIII
   50 – L
   60 – LX
   64 – LXIV
   89 – LXXXIX
   100 – C
   500 – D
   1000 – M

**Challenge 2**
1. 1455
2. 1872
3. 1996
4. 1984
5. £981

**Challenge 3**
1. a) 24
   b) 30
   c) 31
   d) 75
   e) 185
2. a) CX
   b) XCVIII
   c) DCCCIX
   d) CCL
   e) CCLXX

**Pages 18–19**
**Challenge 1**
1. a) 13°C, 8°C, 0°C, −4°C, −8°C, −10°C
   b) 34°C, 28°C, 27°C, 0°C, −1°C, −2°C, −18°C, −19°C
2. a) −8°C
   b) −6°C
   c) −29°C
3. a) 16°C
   b) −1°C
   c) −8°C

**Challenge 2**
1. 15°
2. 11°C
3. 27°
4. 33°
5. 3°C

**Challenge 3**
1. a) Silversty
   b) Sulsbury
   c) 18°
   d) Sulsbury
   e) January and September

**Pages 20–21**
**Challenge 1**
1. 4 correct for 4 marks, e.g. 7, 21, 35, 49, 63, 77
2. 1 and 3
3. 9 + 16 = 25

**Challenge 2**
1. 36 lengths
2. a) 11
   b) 97
3. a) e.g.

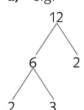

**b)** Variety of possible answers, e.g.

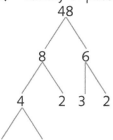

**c)** 1 mark for each tree, e.g.

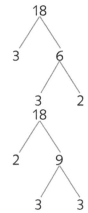

**Challenge 3**
1. a) 100
   b) 400
   c) 16
   d) 1296
2. a) 135
   b) 24
   c) 76
   d) 1728

**Pages 22–23**
**Challenge 1**
1. 45 hours
2. 11
3. 6 football cards, 3 erasers, 5 chocolate coins
**Challenge 2**
1. Answers may vary e.g. Koah has 64 marbles and Zac has 144 marbles
2. 36 books
3. Yes – 64
**Challenge 3**
1. 48 years old or 24 years old
2. Suki
3. a) Mae = 36
      Jake = 17
      Raheed = 27
   b) Mae
4. a) 484: 22 × 22
   b) 496: 4 × 124
   c) 343: $7^3$

**Pages 24–25**
**Challenge 1**
1. a) 7183
   b) 11700
   c) 14689
   d) 18219
2. a) 19712
   b) 53082
   c) 103942
   d) 101764
3. a) 5423
   b) 8989
   c) 17022
   d) 23841
**Challenge 2**
1. a) >
   b) <
   c) <
   d) >
**Challenge 3**
1. a) 70.59
   b) 75.65
   c) 469.31
   d) 700.99
   e) 1268.7

**Pages 26–27**
**Challenge 1**
1. a) 7249
   b) 5353
   c) 1585
   d) 82
2. a) 21025
   b) 29620
   c) 55546
   d) 66401
3. a) 12785
   b) 56980
   c) 84215
   d) 124687
**Challenge 2**
1. a) <
   b) >
   c) >
   d) >
**Challenge 3**
1. 12.01 − 9 = 3.01
   12 − 9.01 = 2.99
   11.99 − 9.02 = 2.97
2. $$\begin{array}{r} 5\,4\,6\,2\,1 \\ -\,2\,3\,8\,3\,1 \\ \hline 3\,0\,7\,9\,0 \end{array}$$

# Answers

## Pages 28–29
### Challenge 1
1. 175 pupils
2. 912 laps
3. 700

### Challenge 2
1. a) Max
   b) 25 370 m
   c) Hannah on Friday
   d) Friday

### Challenge 3
1. a) 1304
   b) 9128
2. 37 540
3. 8424

## Pages 30–31
### Challenge 1
1. a) 112
   b) 228
   c) 896
   d) 2520
   e) 7708
2. a) 357
   b) 352
   c) 3492
   d) 8930
   e) 74 700

### Challenge 2
1. a) 76 × 85 = 6460
   b) 678 × 5 = 3390
2. a) 39 × 37
   b) 84 × 23
   c) 29 × 38
   d) 372 × 18
   e) 846 × 31

### Challenge 3
1. a) 93 364
   b) 9
   c) 1039
   d) 177 345
   e) 9
2. a) 432
   b) 8192
   c) 12 638
   d) 20 768

## Pages 32–33
### Challenge 1
1. a) 123
   b) 384
   c) 781
   d) 4118
   e) 525

2. Answers will vary.
   The quotient must be 124, e.g.
   248 ÷ 2

### Challenge 2
1. a) 2 cakes between 8
   b) 30 biscuits between 10
   c) 24 sweets between 3
   d) 42 marshmallows between 6
   e) 72 jugs of orange juice between 8

### Challenge 3
1. a) 72 ÷ 12
   b) 36 ÷ 6
   c) 81 ÷ 9
   d) 56 ÷ 8
   e) 976 ÷ 4
2. Answers will vary, e.g.
   168 ÷ 6, 280 ÷ 10, 336 ÷ 12
3. Answers will vary, e.g.
   16 ÷ 5, 73 ÷ 9, 145 ÷ 12

## Pages 34–35
### Challenge 1
1. Lissa = 105
   Niall = 82
2. 18 bricks
3. 8

### Challenge 2
1. 5 Popsies and 6 Zoomas
2. 28 cones
3. 24 bats

### Challenge 3
1. 52
2. 13 and a half bananas, 72 strawberries, 18 scoops of ice cream and 2250 ml of milk
3. a) 540 minutes
   b) 1620 minutes
   c) 2430 minutes

## Pages 36–39
### Progress Test 1
1. 40 000
2. 35 200
3. 4.68, 1.2
4. £27
5. 28
6. a) £26
   b) £9
   c) £104
7. 16
8. 105 cupcakes shared between 15
9. >
10. 1380
11. 1936
12. 34 779

13. 81 and 49
14. 29°C
15. 29.6
16. 729
17. 112 781
18. <
19. 1799
20. 48°C
21. e.g.

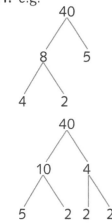

22. 1512
23. 84 × 272
24. 720
25. 45 972

**Pages 40–41**
**Challenge 1**
1. $\frac{3}{4}$
2. $\frac{1}{5}$
3. $\frac{1}{3}$
4. 6 squares shaded
5. 4 squares shaded
**Challenge 2**
1. $\frac{3}{9}$  $\frac{2}{6}$  $\frac{10}{30}$  $\frac{6}{18}$
2. e.g. $\frac{4}{6}, \frac{8}{12}, \frac{20}{30}, \frac{12}{18}$
3. e.g. $\frac{2}{10}, \frac{20}{100}, \frac{5}{25}, \frac{8}{40}$
**Challenge 3**
1. a) $\frac{3}{8}$     b) $\frac{8}{50}$
   c) $\frac{6}{26}$     d) $\frac{9}{27}$

**Pages 42–43**
**Challenge 1**
1. $4\frac{3}{5} - \frac{23}{5}$
   $8\frac{2}{10} - \frac{82}{10}$
   $\frac{7}{3} - 2\frac{1}{3}$
   $\frac{8}{4} - 2$
   $6\frac{4}{5} - \frac{34}{5}$
   $\frac{10}{4} - 2\frac{2}{4}$
   $\frac{74}{9} - 8\frac{2}{9}$

**Challenge 2**
1. a) $\frac{9}{5}$
   b) $\frac{28}{10}$
   c) $\frac{41}{9}$
   d) $\frac{17}{2}$
   e) $\frac{40}{6}$
   f) $\frac{155}{12}$
2. a) $1\frac{1}{3}$
   b) $1\frac{5}{10} = 1\frac{1}{2}$
   c) $2\frac{4}{10} = 2\frac{2}{5}$
   d) $6\frac{1}{8}$
   e) $5\frac{5}{6}$
   f) $9\frac{8}{9}$
**Challenge 3**
1. $32\frac{2}{9}$
2. $\frac{403}{6}$

**Pages 44–45**
**Challenge 1**
1. a) <         b) <
   c) <         d) >
**Challenge 2**
1. $\frac{1}{10}, \frac{1}{6}, \frac{1}{5}, \frac{1}{4}, \frac{1}{3}, \frac{1}{2}$
2. $5\frac{2}{26}, 5\frac{2}{10}, 5\frac{1}{4}, 5\frac{2}{6}, 5\frac{1}{2}$
3. $10\frac{12}{12}, 10\frac{5}{6}, 10\frac{3}{6}, 10\frac{4}{9}, 10\frac{1}{3}$
**Challenge 3**
1. a) $\frac{4}{8}$
   b) $\frac{5}{6}$
   c) $\frac{3}{10}$
   d) $\frac{16}{20}$
   e) $\frac{1}{3}$
2. a) $\frac{4}{10}$
   b) $1\frac{2}{6}$
   c) $\frac{10}{5}$
   d) $\frac{9}{4}$
   e) $\frac{12}{3}$

**Pages 46–47**
**Challenge 1**
1. a) $\frac{2}{4}$ or $\frac{1}{2}$
   b) $\frac{5}{5}$ or 1
   c) $\frac{9}{10}$
   d) $\frac{7}{12}$
   e) $1\frac{3}{4}$
   f) $\frac{1}{6}$

# Answers

g) $\frac{7}{12}$

h) $\frac{1}{8}$

**Challenge 2**

1. a) $\frac{3}{3}$

   b) $\frac{16}{8}$

   c) $\frac{9}{4}$

   d) $\frac{7}{5}$

   e) $\frac{15}{8}$

   f) $\frac{30}{10}$

   g) $\frac{22}{7}$

   h) $\frac{105}{6}$

**Challenge 3**

1. a) $1\frac{2}{5}$

   b) $1\frac{7}{10}$

   c) $\frac{2}{20}$ or $\frac{1}{10}$

   d) $3\frac{1}{7}$

   e) $1\frac{1}{6}$

   f) $3\frac{1}{8}$

   g) $5\frac{3}{7}$

   h) $7\frac{3}{4}$

**Pages 48–49**
**Challenge 1**

1. a) 7

   b) 27

   c) 34

   d) 103

   e) 5220

   f) 39491

   g) 54947

   h) 739944

   i) 932556

**Challenge 2**

1. a) 16.8

   b) 29.5

   c) 73.6

   d) 283.5

   e) 1304.6

   f) 49024.9

**Challenge 3**

1. D, 8.43 because all the others round to 9 when rounded to the nearest whole number.

2. a) 7.3

   b) 9.2

   c) 83.3

   d) 273.5

   e) 827.3

   f) 2358.6

**Pages 50–51**
**Challenge 1**

1. $0.5 - \frac{1}{2}$

   $0.75 - \frac{3}{4}$

   $0.8 - \frac{4}{5}$

   $0.12 - \frac{3}{25}$

   $0.35 - \frac{7}{20}$

   $0.63 - \frac{63}{100}$

   $0.9 - \frac{9}{10}$

**Challenge 2**

1. a) 0.25

   b) 0.75

   c) 1.0

   d) 0.67

   e) 0.4

   f) 0.5

2. a) 0.25

   b) 0.2

   c) 1.0

**Challenge 3**

1. a) $0.51 > \frac{1}{2}$

   b) $0.7 < \frac{8}{10}$

   c) $3.75 > 3\frac{2}{4}$

   d) $0.5 < \frac{6}{10}$

   e) $0.84 > \frac{8}{10}$

   f) $2.39 > 2\frac{3}{10}$

2. a) Answers will vary, e.g. 0.1

   b) Answers will vary, e.g. 0.25

   c) Answers will vary, e.g. 0.5

**Pages 52–53**
**Challenge 1**

1. $\frac{3}{10} - 30\%$

   $\frac{3}{50} - 6\%$

   $0.6 - 60\%$

   $0.2 - 20\%$

   $\frac{10}{10} - 100\%$

   $0.01 - 1\%$

   $85\% - \frac{85}{100}$

   $0.4 - \frac{2}{5}$

**Challenge 2**

1. a) 60%

   b) $\frac{1}{4}$

   c) 99%

   d) 10%

   e) $\frac{6}{8}$

   f) $\frac{2}{5}$

**Challenge 3**

1. a) 50% of £250
   b) 0.1 of £80
   c) $\frac{4}{5}$ of £700
   d) 0.4 of £860
   e) $\frac{2}{5}$ of £1450
   f) 0.5 of £150800

2. $\frac{9}{10}$ off £9156 ✓

**Pages 54–55**
**Challenge 1**

1. a) $\frac{2}{3}$
   b) 2 or $\frac{8}{4}$
   c) 1 or $\frac{8}{8}$
   d) 3 or $\frac{15}{5}$
   e) 6 or $\frac{60}{10}$
   f) 8 or $\frac{40}{5}$
   g) 8 or $\frac{72}{9}$

**Challenge 2**

1. a) $\frac{55}{4} = 13\frac{3}{4}$
   b) $\frac{57}{6} = 9\frac{3}{6} = 9\frac{1}{2}$
   c) $\frac{196}{5} = 39\frac{1}{5}$
   d) $\frac{102}{2} = 51$
   e) $\frac{40}{3} = 13\frac{1}{3}$

**Challenge 3**

1. a) <
   b) <
   c) >
   d) <
   e) >
   f) <

**Pages 56–57**
**Challenge 1**

1. 208 stickers
2. 40%

**Challenge 2**

1. £25.16
2. £211.50

**Challenge 3**

1. TVs R Us have $\frac{3}{10}$ off; Electronic Deals have $\frac{1}{4}$ off
2. a) 135 cm
   b) 10 pupils

**Pages 58–61**
**Progress Test 2**

1. 8879
2. $\frac{1}{4}$ or $\frac{10}{40}$

3. $\frac{22}{8}$
4. $\frac{9}{12}$, $\frac{6}{8}$, $\frac{30}{40}$, $\frac{27}{36}$
5. e.g. $\frac{4}{10}$, $\frac{8}{20}$, $\frac{6}{15}$
6. $\frac{4}{16}$, $\frac{4}{8}$, $\frac{3}{4}$, $\frac{10}{12}$, $\frac{6}{6}$
7. a) 74860
   b) 74900
8. $\frac{9}{12} = \frac{3}{4}$
9. $8\frac{2}{9}$
10. $\frac{37}{12} = 3\frac{1}{12}$
11. $\frac{4}{5}$
12. $\frac{80}{4} = 20$
13. 5874, 924; rule is subtract 2475
14. Shelby's
15. 0.4
16. e.g. $\frac{2}{8}$, $\frac{3}{12}$, $\frac{4}{16}$
17. 136
18. 632.9
19. $\frac{3}{2}$ (or $1\frac{1}{2}$)
20. 60% of 90
21. 60% of 310, 0.5 of 364, $\frac{1}{5}$ of 886, 25% of 660
22. 8 squares should be shaded
23. 30000
24. $\frac{1}{5}$
25. 16
26. 1997
27. 500000
28. 34272
29. 116919
30. 210
31. Variety of possible answers, e.g.

32. 15°C

**Pages 62–63**
**Challenge 1**

1. a) 400 cm
   b) 650 cm
   c) 1000 cm
   d) 2025 cm
   e) 500000 cm
   f) 1075000 cm

# Answers

2.  a)  2000 ml
    b)  600 g
    c)  3.1 km
    d)  0.74 m
    e)  5.25 l
    f)  6.11 kg
**Challenge 2**
1.  a)  <
    b)  >
    c)  >
    d)  >
    e)  <
**Challenge 3**
1.  990 000 g
2.  8800 ml, 8 l, 1.80 l, 880 ml, 0.8 l, 80 ml

**Pages 64–65**
**Challenge 1**
1.  e.g.

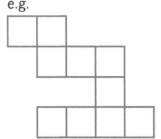

2.  e.g.

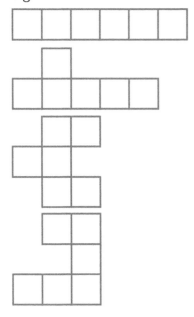

**Challenge 2**
1.  e.g. 30 m × 10 m
2.  e.g. 12 m × 4.5 m

**Challenge 3**
1.  a)  Bluebells – field B
    b)  Lavender – field A
    c)  Gladioli – field C

**Pages 66–67**
**Challenge 1**
1   a)  48 cm²
    b)  132 cm²
    c)  63 cm²
**Challenge 2**
1.  a)  e.g. A rectangle 4 cm × 3 cm
    b)  e.g. A rectangle 6 cm × 3 cm
2.  a)  e.g. 4 m × 5 m
    b)  e.g. 9 m × 8 m
**Challenge 3**
1.  16 tins
2.  12 rolls

**Pages 68–69**
**Challenge 1**
1.  Analogue: 4.38am or 4.38pm
    24-hour 04:38 or 16:38
2.  Analogue: 11:57am or 11:57pm
    24-hour 11:57 or 23:57
3.

**Challenge 2**
1.  a)  >
    b)  >
    c)  >
    d)  >
    e)  <
    f)  <
2.  a)  4 hours 47 mins
    b)  13 hours 46 mins
**Challenge 3**
1.  a)  Bus 1
    b)  Bus 2
    c)  Bus 3
    d)  Bus 3
    e)  Buses 1 and 4

**Pages 70–71**
**Challenge 1**
1.  Friday 20th May at 17:15 (5:15 p.m.)
    1 mark each for correct day/date/time
2.  60 parcels

## Challenge 2
1. 7:40 p.m.
2. 2:30 p.m. or 14:30

## Challenge 3
1. 107.5 or $107\frac{1}{2}$ hours
2. a) 3 hours
   b) 12 hours
   c) 72 hours

## Pages 72–73
### Challenge 1
1. a) £883
   b) £8830
   c) £883
   d) £88.30

### Challenge 2
1. a) £22.90
   b) £42.45 or £37.70 using a family ticket and buying singles
   c) £4.75
   d) £16.35

### Challenge 3
1. a) 9p     b) £3.84
2. a) £39.10     b) 80 laps

## Pages 74–75
### Challenge 1
1. a) 16–24 cm³    b) 60–80 cm³
   c) 168–280 cm³    d) 64–135 cm³

### Challenge 2
1. a) 8 cm³
   b) 8 cm³
   c) 8 cm³
   d) 11 cm³

### Challenge 3
1. a) e.g. 5 × 2 × 2 cm
   b) e.g. 5 × 2 × 3 cm
   c) e.g. 10 × 5 × 2 cm

## Pages 76–77
### Challenge 1
1. **Volume**
   pints, fluid ounces, gallons
   **Distance**
   miles, feet, inches, yards
   **Weight**
   stones, pounds, ounces

### Challenge 2
1. a) 72     b) 228
   c) 264     d) 1536
   e) 28 848
2. a) 216     b) 1440
   c) 3348     d) 12 564
   e) 209 520

## Challenge 3
1. e.g. Height – 51 inches
   Foot – 12 inches
   Hand span – 7 inches

## Pages 78–79
### Challenge 1
1. £702
2. a) 4 packs
   b) 20 packs

### Challenge 2
1. 176 oz
2. 7 tins of paint. (1 mark for working out the missing side length and 1 mark for the correct number of tins.)

### Challenge 3
1. 13
2. 10 rolls. (2 marks for calculating the surface area of each box and 1 mark for the number of rolls.)

## Pages 80–83
### Progress Test 3
1. 4 oz
2. a) 357 m²     b) 76 m
3. 11:00 a.m.
4. 160–270 cm³
5. e.g. 9 m × 12 m
6. 20:26
7. £3.20
8. 13 tins
9. 20 small cakes
10. 1152 inches
11. e.g. 12 m × 9 m
12. 10 136
13. 1594
14. <
15. <
16. 26
17. e.g.

18. Clock hands should show the time 10:40
19. $4\frac{5}{6}$
20. 88 946
21. £17
22. 2021
23. 336 inches
24. £37.44
25. 69 230
26. 19:20
27. $\frac{12}{5}$ (or $2\frac{2}{5}$)

# Answers

## Pages 84–85
### Challenge 1
1. **Regular:**

**Irregular:**

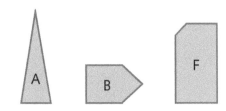

### Challenge 2
1. e.g.

| Name | Regular | Irregular |
|---|---|---|
| Triangle | | |
| Quadrilateral | | |
| Pentagon | | |
| Hexagon | | |
| Octagon | | |

### Challenge 3
1. 19.6 cm
2. 25.8 m
3. 5.5 cm

## Pages 86–87
### Challenge 1
1. 1: Rhombus

2: Trapezium

3: Right-angled triangle

4: Hexagon

5: Rectangle

6: Irregular hexagon

### Challenge 2
1. 8
2. 5
3. 2

### Challenge 3
1. a) Circle
   b) Rectangle
   c) Octagon
   d) Octagon

## Pages 88–89
### Challenge 1
1. A – 3
   B – 1
   C – 7
   D – 2
   E – 5
   F – 4
   G – 6

### Challenge 2
1. a) Cone
   b) Cylinder
   c) Square-based pyramid
   d) Hexagonal prism

### Challenge 3
1.

| Number of edges | Number of faces | Number of vertices |
|---|---|---|
| 18 | 8 | 12 |

2.

| Number of edges | Number of faces | Number of vertices |
|---|---|---|
| 12 | 8 | 6 |

**3.**

| Number of edges | Number of faces | Number of vertices |
|---|---|---|
| 15 | 7 | 10 |

**Pages 90–91**
**Challenge 1**
1. a) False
   b) True
   c) True
   d) True
   e) False
   f) True
   g) False
   h) False
   i) True

**Challenge 2**
1. 6 m × 5 m
2. Any suitable answer, e.g. 12 cm × 2 cm

**Challenge 3**
1. a) 5.5 m and 4.5 m (1 mark for each correct length)
   b) 35 m
   c) 54.5 m²
2. Area = 86 m²          Perimeter = 40 m

**Pages 92–93**
**Challenge 1**
1. a) Obtuse approx. 140°
   b) Reflex approx. 240°
   c) Acute approx. 20°

**Challenge 2**
1.

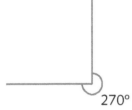

270°

2.

180°

3.

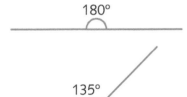

135°

**Challenge 3**
1. 1 mark for the correct shape and 1 mark for the correct angle types, e.g.

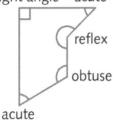

right angle    acute

reflex

obtuse

acute

2. 1 mark for the correct shape and 1 mark for the correct angle types, e.g.

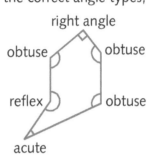

right angle

obtuse          obtuse

reflex          obtuse

acute

**Pages 94–95**
**Challenge 1**
1. 360°
2. 72°
3. 900°

**Challenge 2**
1. a) 60°
   b) 90°
   c) 108°
   d) 120°

**Challenge 3**
1. a) 80°
   b) 205°
   c) 32°

**Pages 96–97**
**Challenge 1**
1. a) (2, 7)
   b) (7, 8)

# Answers

**2. a)**

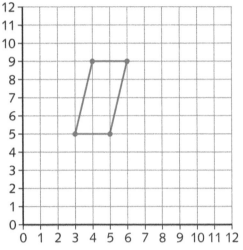

**b)** Parallelogram

**Challenge 2**
1. **a)** (2, 4)
   **b)** (4, 5)
2. (4, 8)

**Challenge 3**
1. **a)** (9, 5)
   **b)** (7, 8)          **c)** (9, 16)
2. 12 cm

## Pages 98–99
**Challenge 1**
1. **a)** Y(6, 8) Z(9, 8)   **b)** M(3, 7) N(5, 7)

**Challenge 2**
1. **a)** T(12, 11) U(14, 7)
   **b)** S(14, 3) U(17, 2)
   **c)** S(0, 2) T(1, 5)

**Challenge 3**
1. (3, 6) (1, 9) (4, 9) (6, 6)
2. (11, 13) (12, 11) (8, 11)
3. (1, 2) (3, 2) (1, 5) (3, 5)

## Pages 100–101
**Challenge 1**
1. **a)** (4, 6) (4, 2) (2, 4)
   **b)** (5, 10) (8, 10) (7, 8) (10, 8)
2. **a)** (2, 8) (4, 8) (3, 7) (1, 7)
   **b)** (5, 3) (7, 3) (6, 2) (4, 2)

**Challenge 2**
1. **a)** (4, 8) (2, 8) (3, 7) (1, 7)
   **b)** (7, 3) (5, 3) (6, 2) (4, 2)

**Challenge 3**
1. **a)** (8, 5)
   **b)** (7, 2) (4, 3) (3, 6) (6, 5)

## Pages 102–103
**Challenge 1**
1. **a)** Yellow House   **b)** 18 points
   **c)** 2014

**Challenge 2**
1. **a)** Spain        **b)** America
   **c)** 2.18 m

**Challenge 3**
1. **a)**

| Class | Number of pages read | Total pages read |
|---|---|---|
| 5HY | 𝍷𝍷𝍷𝍷𝍷𝍷𝍷 𝍷𝍷𝍷𝍷𝍷𝍷𝍷III | 68 |
| 5AS | 𝍷𝍷𝍷𝍷𝍷𝍷𝍷 𝍷𝍷𝍷𝍷𝍷𝍷𝍷 𝍷𝍷I | 81 |
| 5LS | 𝍷𝍷𝍷𝍷𝍷𝍷𝍷 𝍷𝍷𝍷𝍷𝍷𝍷𝍷 𝍷𝍷𝍷I | 86 |
| 5LM | 𝍷𝍷𝍷𝍷𝍷𝍷 𝍷𝍷𝍷𝍷III | 53 |
| 6JS | 𝍷𝍷𝍷𝍷𝍷𝍷 𝍷𝍷𝍷𝍷𝍷𝍷 𝍷𝍷IIII | 74 |
| 6PD | 𝍷𝍷𝍷𝍷𝍷𝍷 𝍷𝍷𝍷𝍷𝍷𝍷 𝍷𝍷𝍷𝍷III | 83 |
| 6KF | 𝍷𝍷𝍷𝍷𝍷𝍷 𝍷𝍷𝍷𝍷𝍷𝍷 𝍷𝍷𝍷IIII | 79 |
| 6SH | 𝍷𝍷𝍷𝍷𝍷𝍷 𝍷𝍷𝍷𝍷𝍷𝍷 𝍷𝍷𝍷𝍷II | 82 |

**b)** 5LS          **c)** 606 pages
**d)** 30 pages

## Pages 104–105
**Challenge 1**
1.

| Zone | Number of people at different times of day | | | | |
|---|---|---|---|---|---|
| | 10 a.m. | 12 noon | 2 p.m. | 4 p.m. | 6 p.m. |
| Play Zone | 190 | 65 | 255 | 240 | 55 |
| Relaxing River | 90 | 105 | 120 | 200 | 75 |
| Heights | 150 | 90 | 225 | 195 | 65 |
| Terror Fall | 50 | 30 | 205 | 185 | 10 |

**Challenge 2**
1.  **a)** Play Zone
    **b)** Play Zone
    **c)** 325 people
**Challenge 3**
1.  **a)** 6 p.m.
    **b)** e.g. meal time; people will have left the park; nearly closing time
2.  **a)** Terror Fall
    **b)** e.g. the rides are not suitable for the whole family; age/height restrictions

**Pages 106–107**
**Challenge 1**
1.  A – 3
    B – 1
    C – 2
**Challenge 2**
1.  1 mark for each of the four parts of the journey depicted on a graph, e.g.

**Challenge 3**
1.  **a)** 1 mark each for the following: numbers labelled on both axes; axis titles; points plotted; points joined with a smooth line

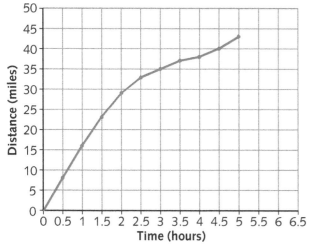

    **b)** 48–52 miles

**Pages 108–111**
**Progress Test 4**
1.  37
2.  Any angle between 90° and 180°

3.  **a)** (9, 10) (11, 10) (10, 7)
    **b)** (13, 6) (15, 6) (14, 3)
4.  e.g.

5.  Between 180° and 360°
6.  2 circles and 1 rectangle
7.  (9, 7)
8.  **a)** £474.11
    **b)** £130.40
    **c)** £18
9.  e.g.

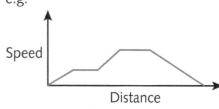

10.

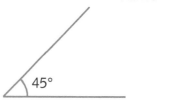

11. 1 mark for each of the four types of angle, e.g.

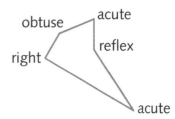

12. 45°
13. 30°
14. 0.25 and $\frac{1}{4}$ (or $\frac{25}{100}$)
15. 138 seconds or 2 minutes 18 seconds
16. 1 mark for correct shapes and 1 mark for correct number: 6 rectangles and 2 hexagons
17. 106 825
18. 9 edges
19. **a)** 126 m²
    **b)** 46 m
20. $\frac{32}{12}$ or $\frac{8}{3}$ or $2\frac{2}{3}$
21. 36 cm

# Notes

# Progress Test Charts

## Progress Test 1

| Q | Topic | ✓ or ✗ | See page |
|---|-------|--------|----------|
| 1 | Place Value | | 12 |
| 2 | Rounding Numbers | | 14 |
| 3 | Subtraction | | 26 |
| 4 | Word Problems | | 22 |
| 5 | Properties of Numbers | | 20 |
| 6 | Rounding Numbers | | 14 |
| 7 | Properties of Numbers | | 20 |
| 8 | Division | | 32 |
| 9 | Place Value | | 12 |
| 10 | Division | | 32 |
| 11 | Roman Numerals | | 16 |
| 12 | Subtraction | | 26 |
| 13 | Properties of Numbers | | 20 |
| 14 | Negative Numbers | | 18 |
| 15 | Multiplication | | 30 |
| 16 | Multiplication | | 30 |
| 17 | Subtraction | | 26 |
| 18 | Place Value | | 12 |
| 19 | Roman Numerals | | 16 |
| 20 | Negative Numbers | | 18 |
| 21 | Properties of Numbers | | 20 |
| 22 | Addition | | 24 |
| 23 | Multiplication | | 30 |
| 24 | Word Problems | | 34 |
| 25 | Addition | | 24 |

## Progress Test 2

| Q | Topic | ✓ or ✗ | See page |
|---|-------|--------|----------|
| 1 | Subtraction | | 26 |
| 2 | Comparing and Ordering Fractions | | 44 |
| 3 | Improper Fractions | | 42 |
| 4 | Equivalent Fractions | | 40 |
| 5 | Equivalent Fractions | | 40 |
| 6 | Comparing and Ordering Fractions | | 44 |
| 7 | Rounding Numbers | | 14 |
| 8 | Adding Fractions | | 46 |
| 9 | Adding Fractions | | 46 |
| 10 | Improper Fractions | | 42 |
| 11 | Subtracting Fractions | | 46 |
| 12 | Multiplying Fractions | | 54 |
| 13 | Subtraction | | 26 |
| 14 | Fraction, Decimal and Percentage Equivalents | | 52 |
| 15 | Fraction, Decimal and Percentage Equivalents | | 52 |
| 16 | Equivalent Fractions | | 40 |
| 17 | Rounding Decimals | | 48 |
| 18 | Rounding Decimals | | 48 |
| 19 | Fraction, Decimal and Percentage Equivalents | | 52 |
| 20 | Fraction, Decimal and Percentage Equivalents | | 52 |
| 21 | Fraction, Decimal and Percentage Equivalents | | 52 |
| 22 | Equivalent Fractions | | 40 |
| 23 | Place Value | | 12 |
| 24 | Fraction, Decimal and Percentage Equivalents | | 52 |
| 25 | Multiplying Fractions | | 54 |
| 26 | Roman Numerals | | 16 |
| 27 | Place Value | | 12 |
| 28 | Multiplication | | 30 |
| 29 | Subtraction | | 26 |
| 30 | Rounding Numbers | | 14 |
| 31 | Properties of Numbers | | 20 |
| 32 | Subtraction | | 26 |

# Progress Test Charts

## Progress Test 3

| Q | Topic | ✓ or ✗ | See page |
|---|-------|--------|----------|
| 1 | Equivalent and Imperial Units of Measure | | 76 |
| 2 | Perimeter/Area | | 64/66 |
| 3 | Time Word Problems | | 70 |
| 4 | Volume | | 74 |
| 5 | Area | | 66 |
| 6 | Time | | 68 |
| 7 | Money | | 72 |
| 8 | Measurement Word Problems | | 78 |
| 9 | Money | | 72 |
| 10 | Equivalent and Imperial Units of Measure | | 76 |
| 11 | Perimeter | | 64 |
| 12 | Multiplication | | 30 |
| 13 | Roman Numerals | | 16 |
| 14 | Equivalent Fractions | | 40 |
| 15 | Time | | 68 |
| 16 | Time Word Problems | | 70 |
| 17 | Perimeter | | 64 |
| 18 | Time | | 68 |
| 19 | Improper Fractions | | 42 |
| 20 | Multiplication | | 30 |
| 21 | Money | | 72 |
| 22 | Roman Numerals | | 16 |
| 23 | Equivalent and Imperial Units of Measure | | 76 |
| 24 | Money | | 72 |
| 25 | Multiplication | | 30 |
| 26 | Time | | 68 |
| 27 | Multiplying Fractions | | 54 |

## Progress Test 4

| Q | Topic | ✓ or ✗ | See page |
|---|-------|--------|----------|
| 1 | Rounding Decimals | | 48 |
| 2 | Different Types of Angle | | 92 |
| 3 | Translation/Reflection | | 98/100 |
| 4 | 2-D Shapes | | 86 |
| 5 | Different Types of Angle | | 92 |
| 6 | 2-D Shapes | | 86 |
| 7 | Co-ordinates | | 96 |
| 8 | Tables and Timetables | | 102 |
| 9 | Line Graphs | | 106 |
| 10 | Calculating Angles | | 94 |
| 11 | Different Types of Angle | | 92 |
| 12 | Calculating Angles | | 94 |
| 13 | Calculating Angles | | 94 |
| 14 | Fraction, Decimal and Percentage Equivalents | | 52 |
| 15 | Time Word Problems | | 70 |
| 16 | 2-D Shapes/3-D Shapes | | 86/88 |
| 17 | Multiplication | | 30 |
| 18 | 3-D Shapes | | 88 |
| 19 | Perimeter/Area | | 64/66 |
| 20 | Multiplying Fractions | | 54 |
| 21 | Perimeter | | 64 |

What am I doing well in? _____

_____

What do I need to improve? _____

_____